HIGH TEMPERATURE DRYING

Also published by Miller Freeman Publications for the forest products industry

Computer Control Systems for Log Processing and Lumber Manufacturing
by Ed M. Williston

Dry Land Log Handling and Sorting
by Charles M. Hampston

Farming the Small Forest
by Laurence C. Walker

The Logging Business Management Handbook
by Ronald R. Macklin

Logging Practices, revised edition
by Steven Conway

Lumber Manufacturing: The Design and Operation of Sawmills and Planer Mills, revised edition
by Ed M. Williston

Managing Mill Maintenance: The Emerging Realities
by Richard F. Baldwin

Modern Particleboard and Dry-process Fiberboard Manufacturing
by Thomas M. Maloney

Operations Management in the Forest Products Industry
by Richard F. Baldwin

Plywood Manufacturing Practices, second edition
by Richard F. Baldwin

Quality Control in Lumber Manufacturing
edited by Terence D. Brown

Saws: Design, Selection, Operation, Maintenance, second edition
by Ed M. Williston

Small Log Sawmills: Profitable Product Selection, Process Design and Operation
by Ed M. Williston

Timber Cutting Practices, third edition
by Steve Conway

Directory of the Forest Products Industry

FOREST INDUSTRIES Mill Map

FOREST INDUSTRIES North American Factbook

Secondary Wood Products Manufacturers Directory

Larry Culpepper

HIGH TEMPERATURE DRYING

Enhancing Kiln Operations

Miller Freeman Publications, Inc.
San Francisco, Atlanta, Boston,
Chicago, New York, Brussels

Library of Congress Catalog Card Number: 90-61153
International Standard Book Number: 0-87930-186-4

For information contact the publishers, Miller Freeman Publications, Inc., 500 Howard Street, San Francisco, California 94105, USA.

Cover design by Tony Agpoon
Printed in the United States of America
First printing, April, 1990

To my brother, Britt,
and his wife, Jo.

If I had known I would spend
as much time with you as I did . . .

Well . . . you know the rest.
Thanks.

Contents

Man's mind, once stretched by a new idea, never regains its original dimensions.

Oliver Wendell Holmes

Introduction

> The secret of happiness is not in doing what one likes, but in liking what one does.
>
> James M. Barrie

This is an exciting time to be involved with kiln drying lumber. The introduction of computerized controls over the past decade has opened up new horizons in drying technology. Not only has this fostered the introduction of new ideas and techniques in drying, but older ideas are being rediscovered and successfully applied for the first time.

When I became interested in high-temperature drying over ten years ago, these kilns were considered necessary evils by many companies. In my travels as an inspector for the Southern Pine Inspection Bureau (SPIB), I was amazed at the lack of attention paid to the drykilns. Poor drying quality was accepted as part of the price mills paid to get their lumber within the moisture provisions of the grading rules. After all, southern pine had a well-deserved reputation for warping if you even looked at it sideways. There wasn't much you could do about the nature of the species.

Rare was the kiln operator who considered kiln drying his career. Most operators accepted their responsibility as a price that had to be paid to get somewhere else. If you did a reasonable job here, you could move into the important jobs at the sawmill or planing mill. Anybody could run the kiln. Few stayed in the position long enough to really learn the intricacies of the trade. But then, who could teach them? The high-temperature drying process had come out of the universities without the base of knowledge required to know how the equipment should be designed, much less perform. The kiln manufacturers were asked to build this drying equipment based on conjecture and theory. Speed alone was the objective. Two kilns built by a single manufacturer six months apart might be very different in performance. No one kept up with the few gains in technology. Quality? Forget it.

As I traveled around it seemed that there had to be a better way. As an SPIB inspector I found it frustrating to deal with mills that had moisture problems but no clear idea how to attack them. In 1980 I left the Bureau and asked Union Camp Corporation to hire me for one year to collect information on these kilns for a master's thesis. In return I would help them with their drying program. Computerized multizone drykilns were born of this research and I spent almost four years with Union Camp. I never did complete my thesis.

There are two driving forces behind my work with drykilns. First, quality is important to me. As an SPIB inspector I learned the details of quality and how it affects the user. Before the Bureau I was a lumber user, working as a carpenter and cabinetmaker, hobbies I continue to enjoy. Second, I love to make things work better, more easily, and more efficiently. No matter what you are doing, there is always a better way. Sometimes all it takes is someone new looking at something fresh or for the first time. Finding that little extra something is a gratifying experience.

This book is about high-temperature drying of southern pine, although many of the principles that apply to this process apply to others as well. Strictly speaking, high-temperature drying is drying done at temperatures in excess of the boiling point of water. The past decade has seen great strides in the potential quality of lumber you can expect from these kilns. The past five years has seen a general heightened appreciation of the true costs of lumber drying as a whole. Everyone seems involved. High-temperature drying of southern pine has finally gained the attention it deserves. I expect future developments to move at a faster pace.

I have attempted to gather up-to-date information on high-temperature drying and present it in an easy-to-understand form for everyone from the stacker to the people in the front office. Most important, it is aimed at the operators. They see it all and are in the best position to judge how different areas affect them. They cannot do it alone. Quality drying is the responsibility of everyone from the sawmill to the person who loads dressed lumber on the truck. All must work together to get the job done. This book will help coordinate efforts toward developing a kiln improvement program for the mill. Everyone benefits from a successful program.

The purpose of this book is to educate new initiates into industry practices and stimulate new ideas. It serves to increase the understanding of high-temperature drying on the operator level and help the operator identify potential problem areas. Finally, it attempts to provide more efficient utilization of our natural resources to increase the amount of usable lumber derived from the timber base. Not coincidentially, this assures the profit and long-term success of any sawmill.

This book is the culmination of ten years' work in the drying of southern pine. There is still much to do: new ideas to try and new kilns to build. The work goes on. Over these years I have met many people whose ideas shaped the final form of this text. There is no way to recall them all, but I would like to thank those who made the greatest impact in bringing this book into being: J. W. Smithey, W. H. Skipworth, Garland Gravely, Terry Brown, George Culp, Cliff McConnell, Martin Nye, Henry Schuette, Jim Teaford, Gayle Bridges, Milton Stahmer, and Bob Smith.

Larry G. Culpepper
September 1989

CHAPTER 1

Lumber Manufacture and Its Impact on Drying

People forget how fast you did a job—but they remember how well you did it.

Howard W. Newton

Why Worry About Lumber Manufacture?

Why should you worry about lumber manufacture? Because poor manufacture costs money. Quality drying is achieved by reducing the variability of factors affecting the drying process. Poor manufacture affects lumber yields, handling costs, energy costs, equipment life, safety, moisture distributions, lumber, and grade recovery by introducing variability. Concern for controlling this begins at the sawmill with the delivery and unloading of logs onto the yard.

Log Storage

The beginning moisture content of green lumber affects its drying rate.[1] All other factors being equal, boards with high moisture content at the beginning of drying will be higher at charge end. Studies have shown that the moisture content of timber varies because of many factors.[2] For southern pine these include:

- species: slash, longleaf, shortleaf, loblolly, and so on.
- tree age
- wood density
- geographic or site location
- height from ground
- heartwood versus sapwood
- storage method

For practical purposes the only factor we can control is storage method. Drying problems related to storage are greatest at mills that store logs under water sprays. Expect trouble when you mix water-stored and fresh-cut logs in the same production run. Wet and overdry spots occur, which you cannot pinpoint to a specific area within the kiln. Extensions

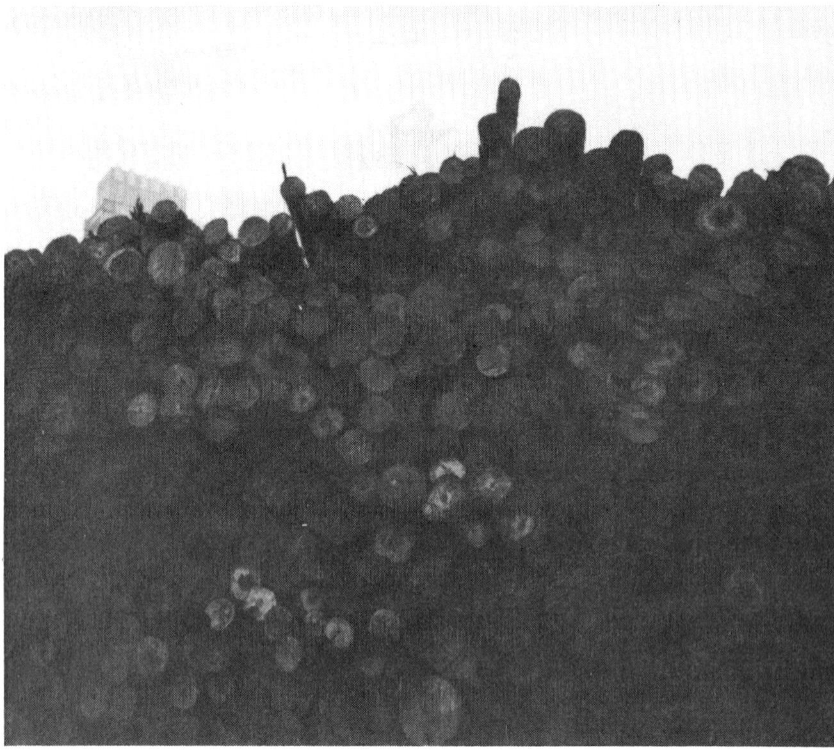

Figure 1.1. Logs held under sprinkler storage. Note mixing of new inventory with old. Logs must be rotated through the system on a first-in, first-out basis to reduce drying problems.

in drying time are common. Lumber from logs held under long-term water storage must be dried separately from other production. To improve the effectiveness of your water-sprinkled log yard, make certain your sprinkler system keeps the ends of logs wet.

Whether or not you store logs under water, you should place them through the mill on a first-in, first-out basis. This reduces moisture variations due to predrying of logs on the yard and helps control the problems associated with insect and fungal attack.

Variations in Lumber Manufacture

Manufacturing all lumber to uniform dimensions eliminates manufacture as a source of drying problems. Unfortunately, lumber size can vary a great deal depending on the setwork accuracy of even a single machine. The variation between different machine centers is often extreme.

Check machine centers daily and make corrections as problems occur. Identify substandard equipment and repair or replace it as needed. The long-term solution to this problem is the effective implementation of a quality control (QC) program in the sawmill. A well run QC program pays back in increased lumber yield through reductions in target size and increased grade recovery. It is well worth the effort for any mill. Guidelines for starting a QC program are beyond the scope of this book. For more information I recommend Terry Brown's text, *Quality Control in Lumber Manufacturing.*[3]

Figure 1.2. Wide variation in board thickness allows thin boards adjacent to thick ones to move at will while drying. Uniform lumber thickness helps keep all lumber flat.

Thickness

Of all the variations in manufacture, variations in lumber thickness have the greatest impact on drying. These variations affect grade and equipment as well as moisture distributions. Thick lumber takes longer to dry than its thinner neighbor, and variations of 0.125 inch are significant. Actual measurement of the impact of thickness variation is difficult given the many variables that affect final moisture content. Computer simulations of drying rates indicate that a ±0.10-inch variation in lumber thickness results in a 3.5 percent variation in final moisture content. This variation adds to the many sources of variation common in any drying system. Tight control of target thickness helps narrow the final moisture distribution of your product.

Mills with poor thickness control experience excessive degrade and high sticker costs as well. The data in Table 1.1 from a mill illustrates how bad things can be.

This mill has serious drying problems. Stick losses are high, as sticks conform to the uneven surface across the width of stacked units. Thin boards side by side with thick boards dry without restraint and are prone to warp, contributing to degrade. Many do not dress at the planing mill. Thick pieces dry more slowly and force you to run kilns longer to assure that all the lumber is dry. They cause excessive machinery wear or require extra handling at the planing mill. Imagine the impact of that 2.577-inch thick board as it enters a planing machine at 1,000 fpm.

Thick lumber is a significant source of loss, yet it pales in comparison to the losses from thin lumber. Unless controlled, the mill in our example must increase target sizes to reduce the 15 percent to 35 percent thin production. This affects the very survival of any mill, because it impacts on lumber recovery factors and grade yield. Although this is an extreme example, it illustrates the potential problems of variations in thickness. The problems do not go away with smaller variations, they simply reduce

Table 1.1 *A Mill with Poor Thickness Control*

		Thickness				
Item	*# Samples*	*Avg.*	*Min.*	*Max.*	*Range*	*% Thin*
2 × 6 16′	48	1.834″	1.423″	2.030″	0.607″	31.3
2 × 6 18′	48	1.754″	1.385″	2.577″	1.192″	35.4
2 × 4 12′	51	1.745″	1.565″	1.914″	0.349″	25.5
2 × 8 12′	32	1.774″	1.575″	1.940″	0.365″	18.8
4 × 4 16′	20	3.841″	3.025″	3.976″	0.951″	15.0

Figure 1.3. Extreme thickness variation causes stick breakage.

in size. Variations in final moisture content continue. Machinery wear and additional handling costs, although difficult to measure, are real. Sticks forced to conform to uneven lumber surfaces break the next time through. Thin boards drying without support are uneven in appearance and often lose grade or must be trimmed after dressing. Thickness control is essential to quality drying.

Sawing taper into lumber is a related thickness problem. A board that is tapered is thicker on one end than on the other. Look for this problem at mills operating conventional or hybrid Chip-n-Saws. Tapers ranging from 0.002- inch to 0.004- inch per foot are common. The thicker ends dry with slightly higher final moisture content. Thin ends are no problem as long as they dress at the planing mill. Tapered lumber is easily spotted in lumber stacked to enter the kiln. One end of the stacked lumber pile is always higher than the other. From a distance the tops of several units in a line have a sawtooth appearance. Stacking 50 layers, each with a taper of 0.004- inch per foot, gives an overall stacked unit height difference of 3.2 inches end to end in 16-foot lumber. Overhead baffles do not close the gaps on this lumber. This results in some air loss over the units and upsets

Figure 1.4. Tapered lumber forces the sticklayers to double up sticks on one end to compensate for the height difference between ends.

airflows. The serious consequences occur when sticklayers notice the lumber surface is uneven at the stacker. Rather than alert supervisors to correct the problem, they often double up sticks or turn them on edge at the low end. This gives them a level surface for laying sticks but does not give full support to the lumber. The lumber warps as it dries.

Occasional variations in thickness from snaking horizontal bandsaws result in the loss of individual boards. These boards may simply be too thin and often cannot be trimmed. When too thick, losses extend to neighboring boards and to stickers. Keep in mind that each snaked board you see has a companion. Correct this problem by examining your saw filing and saw guide maintenance replacement practices.

Figure 1.5. Variation in lumber thickness from snaking bandsaws. When the problem is this severe, significant grade losses occur.

Width

Variations in width within a single size (2 × 4s, 2 × 6s, etc.) have very little impact on drying. The real problem occurs when variations are so great or targets so wide that we must drop a board from the width of the unit.

The decision to drop a board is forced by stacker-width capacity. Full count layers that are too wide to fit between the stacker guides force the mill to drop one board from each unit. Rather than fix the problem by tightening target widths, mills make adjustments farther along in the system. Tightening target sizes pays off in increased kiln capacity and often postpones the need to build more kilns. Increased capacity alone saves an average of one kiln charge every fifteen days. Use it to reduce your drying costs.

Drying rates are affected by the total width of a stacked unit. When stacking to the total 8-foot width all units but the 2 × 10s are a full 96 inches wide. Often mills notice that 2 × 10s mixed in with other items in a kiln are drier. This is due to the reduced wood volume across the unit

Table 1.2 *Comparison of Unit Widths*

8″ Wide Unit	*# Pieces Per Layer*				
	One Board Dropped		*Full Width Capacity*		
Item	*# Pieces Per Layer*	*Total Width*	*# Pieces Per Layer*	*Total Width*	*Increased Capacity*
2 × 4	23	92″	24	96″	4.3%
2 × 6	15	90″	16	96″	6.7%
2 × 8	11	88″	12	96″	9.1%
2 × 10	9	90″	9	90″	0.0%
2 × 12	7	84″	8	96″	14.3%

and is an argument for drying 2 × 10s separately. At mills that have lost one board in width the variation is 8 inches among all items. This can be a significant source of moisture variation when these mills mix widths within a charge.

Length

Poor trimming affects the overall airflow distribution in a kiln by leaving openings through which air bypasses the lumber. This reduces the efficiency of the air delivery system and creates small imbalances in airflow uniformity. Its impact on drying is not easy to measure and is often overlooked. The two trimming problems most often found are variations in length due to trimsaw settings and mistrims.

From any trimsaw, several combinations of saws produce any single length. When the maximum length produced by a mill is 16 feet only one saw combination produces that length. However, 14-foot lumber can be produced by trimming both the near or far end. Trimming 4 feet from either end or 2 feet from both ends produces 12-foot lumber, and so on. If these saws are not set properly the actual lengths of lumber produced from the various saw combinations are not equal. I have seen variations to 1⅝ inches. These variations make it difficult to butt trucks tightly end to end within the kiln. The openings offer an avenue for air to short circuit around the lumber.

The following saw settings are measured from the even end stop on the trimmer. These settings give a maximum range in length for all combinations of ¼ inch. This results in a neat, tight unit going into the kiln. The minimum overlength values do not take into consideration that *these lengths will be shorter by one saw kerf.*

Figure 1.6. The maximum acceptable gap between butted units within the kiln. We would prefer to have no gap, but unless this one aligns with a control sensor, it is hard to measure its impact.

Table 1.3 ***Trimmer Saw Positions (In inches from even end line)***

0′ saw	*2′ saw*	*4′ saw*	*6′ saw*	*8′ saw*	*10′ saw*
0.75″	25.00″	49.25″	73.50″	98.00″	122.25″

12′ saw	*14′ saw*	*16′ saw*	*18′ saw*	*20′ saw*
146.50″	170.75″	195.25″	219.50″	243.75″

Mistrimming and missorting lumber is far more serious. The extra-long lengths stacked into units create large gaps between units. This is a sawmill problem and responsibility for its correction rests there. Repair problems when you find them. Such problems include the failure of saws to drop, motor failure, loss of limit switches, and logic problems. Canadian-type trimmers must run at speeds that permit operators adequate time to position boards before they enter the saws. It is inexcusable for problems to continue without making efforts to correct them. Ignoring them affects the entire drying process and endangers personnel who must load lumber into the kilns.

Airflow is upset by large gaps in the lumber. If mistrimming results in boards extending beyond the length of the units, then large gaps open between adjacent units on a truck. In severe cases this limits the amount of lumber that will fit in the kiln, reducing kiln capacity. Once out of the sawmill the problem can only be corrected by someone with a chainsaw. When loading by forklift individual units can be trimmed at ground level on the stacker rollcase or before setting on the trucks. For transfer operations there is no way personnel can follow prescribed chainsaw safety procedures and trim this lumber safely.

If safety is not reason enough to trim it right in the first place, then consider money. The costs for manpower, fuel, and equipment are obvious. Are you absolutely certain that the end of the board that was trimmed is the end that should have been trimmed? When trimming is this sloppy the stacking crew pays less attention to even-ending lumber on the stacker. The board may not have needed trimming in the first place. It will be trimmed again at the planing mill after adding on drying and handling costs. Did you needlessly lose 2 feet of board? Perhaps not, but you can never be certain. When you consider how much one man can cost a mill with a simple chainsaw, the decision to trim it right at the sawmill is easy.

Table 1.4 ***Resulting Length Combination Comparison Data***

	Maximum Length	*Minimum Length*	*Length Range*	*Nominal Length*	*Minimum Overlength*
8′ combinations	97.25″	97.25″	0.00″	96.00″	1.25″
10′ combinations	121.75″	121.50″	0.25″	120.00″	1.50″
12′ combinations	146.00″	145.75″	0.25″	144.00″	1.75″
14′ combinations	170.25″	170.00″	0.25″	168.00″	2.00″
16′ combinations	194.50″	194.50″	0.00″	192.00″	2.50″
18′ combinations	218.75″	216.75″	0.00″	216.00″	2.75″
20′ combinations	243.00″	243.00″	0.00″	240.00″	3.00″

An unusual trimming problem occurred at a mill that dried studs from two company mills. Those which were shipped in were uniform in length and fully loaded the kilns. The local mill produced untrimmed studs that were 8 inches longer. Since lumber from both mills could not be dried together, the local mill lost one kiln car from each track for a capacity reduction of 17 percent. A kiln charge of local lumber left a 6-foot gap on each track, which the end baffles could not close. This open avenue for air lowered overall air velocity and extended drying time. Moisture distributions were very wide. Determining when to pull the kilns was a guessing game you could not win. This was a design flaw of the plywood plant. There simply was no room for a trimmer.

Mismanufacture

Mismanufacture is any deviation from established target size. In practice, however, most people take this term to mean that the lumber has an unnatural defect manufactured into it. Snaking saws produce a type of mismanufacture we have already discussed. Skipped lumber resulting from mismatched saws in horizontal or vertical double arbor machine centers is another. Rabbeted edges from Chip-n-Saws and related machinery are particularly noticeable. These defects do not affect drying unless they are serious enough to alter the drying speed at the defect or damage lumber in their vicinity. The most serious considerations with these defects are those of appearance, grade, and volume yield.

The grading rules establish the amount of each defect permissible in each grade. Finish and shop grades are graded on appearance or suitability for remanufacturing. They exclude these types of defects. Structural grades are judged by strength and often permit a significant amount.[4]

Figure 1.7. The result of serious mistrimming at the sawmill. Is he trimming the end of the board he should be trimming?

For example, rabbeted edge is graded according to wane rules. On a #2 2 × 4 it may extend 1/2 inch deep and a little over 1 1/8 inches wide for the full length of the board. In 5 percent of pieces it may extend 1/2 inch deep and 1 3/4 inches wide or 1 inch deep and 7/8 inch wide for one-

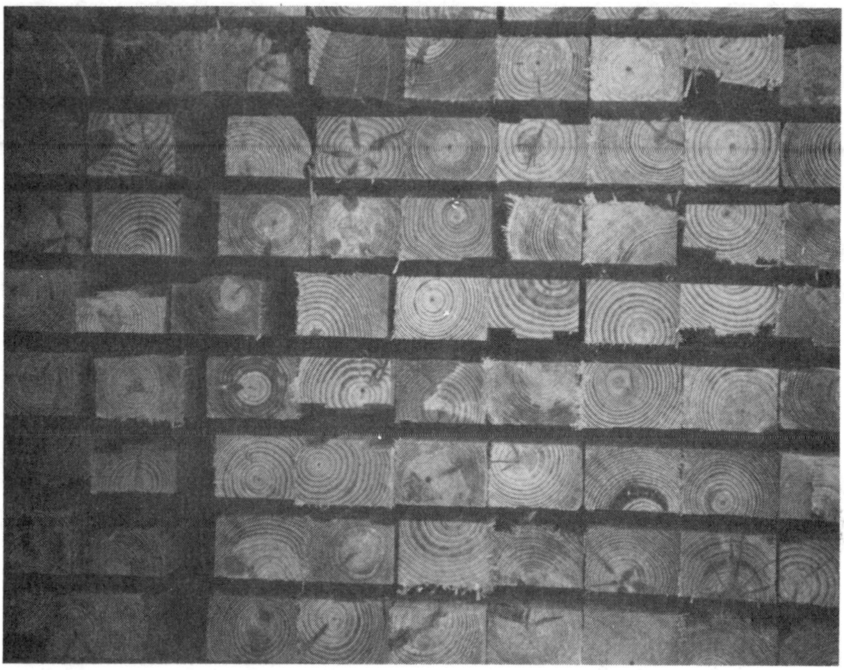

Figure 1.8. Mismanufacture from a Chip-n-Saw. Severe rabbeted edge increases trim and grade loss for this mill. This is a sawmill maintenance problem.

quarter the length. This material is difficult to sell in today's marketplace. You cannot afford to overlook the importance of appearance and its impact on the customer.

It pays to take the position that you will not sell anything that you yourself would not buy. When you shop at a building supply store for that weekend project, you are spending your money. That porch, deck, or shed is a reflection of your pride in ownership. If you are like me, you want the best you can afford. Appearance means a great deal in these decisions. The people who purchase your lumber feel the same way.

As a quality supervisor for the Southern Pine Inspection Bureau (SPIB) I inspected a carload of 2 × 4s delivered to a truss plant. This customer was proud of his trusses and objected to the skipped appearance of the lumber. Almost every board showed skipped dressing from mismatched saws. The lumber had been dried in a wood-fired direct-fired kiln and the skips stood out under the yellow millbright coating. The load looked pretty rough, but the skips were not deep and were well within the rules. Five percent of a shipment may be below grade and, after

Figure 1.9. Fuzzy edge grain resulting from dull saws. These fibers slow air movement through the lumber.

inspecting a few bundles, I found the total degrade too close to recommend a complete claim inspection. The customer still did not want the lumber.

During a break I had the opportunity to watch the plant in action. Lumber from one supplier was breaking up as it passed through the truss saws. My suspicions aroused, I inspected a full bundle on the yard and found it 13 percent below grade, mostly due to crossbreaks. When I presented this information to the plant manager, he replied that this lumber was from his best supplier. It was always clean and good looking with little wane or warp. He paid a premium price and was glad to have it on his yard. He would not call a claim inspection. Appearance is important to your customers, too. Many will pay extra for it.

Mismanufacture that affects drying is rare and usually involves the surface quality of the lumber. When saws are not maintained or properly filed, a fuzzy surface results. This is a particular concern to mills processing steamed peeler cores from plywood plants. Sash gang saws tear grain at board edges. Both problems slow airflow through the stickers, extend drying time and contribute to wide moisture distributions. I have seen situations where you could not see through stick openings from one side to the other due to the fuzzy surface.

The quality of manufacture from any sawmill is a statement of the pride and responsibility of its employees. It cannot be shifted down the line to someone else. It affects the total productivity for the mill. Logyard, sawmill, stacker, drykilns, and planing mill all must work together to identify and solve problems. Everything you do—or don't do—has an impact farther along in the system. It is always easier to do a job right the first time than correct the problems of poor workmanship.

CHAPTER GUIDELINES FOR REDUCING VARIABILITY IN DRYING

1. Rotate log inventory through the sawmill on a first-in, first-out basis to reduce predrying of logs on yard.
2. Dry lumber manufactured from water-stored logs separate from other production.
3. Implement a quality control (QC) program in the sawmill to reduce and maintain target sizes for thickness, width, and length.
4. Notify the people responsible when poor manufacture is observed. Repair the cause immediately.
5. Hold the sawmill personnel accountable for the quality of the lumber. This can be done by deducting plantwide waste from the sawmill cut.

REFERENCES

1. Denig, Joseph, Stephen J. Hanover, and C. Arthur Hart. *Kiln Drying Southern Yellow Pine Lumber.* The North Carolina Agricultural Extension Service. Publication No. AG-402, 1988.
2. Koch, Peter. *Utilization of the Southern Pines*. Agriculture Handbook No. 420, U.S. Department of Agriculture, Forest Service, 1972.
3. Brown, Terence D. *Quality Control in Lumber Manufacturing*. San Francisco: Miller Freeman Publications, 1982.
4. Southern Pine Inspection Bureau. *Grading Rules—1977 Edition*. Pensacola, FL: Southern Pine Inspection Bureau, 1977.

CHAPTER 2

Sorting and Green Lumber Handling

Nothing in the world can take the place of persistence. Talent will not; nothing is more common than unsuccessful men with talent. Genius will not; unrewarded genius is almost a proverb. Education will not; the world is full of educated derelicts.

Persistence and determination alone are omnipotent.

Calvin Coolidge

Why Sort Your Lumber?

The primary purpose of sorting is to separate lumber items for sale or further processing. Sorting makes better use of drying capacity by allowing you to pack more lumber into your kilns. It improves drying quality. It is rare to find a mill that does not sort directly out of the sawmill. Each thickness is sorted for length and width, then stacked and dried in charges of one thickness. This is the recommended practice at all sawmills.

A few years back you could find mills that mixed all widths and lengths of a thickness into boxpiled packages for drying. The dried lumber was then sorted in a "rip mill" by width, length, and grade. Rip mills were equipped to upgrade lumber by ripping or trimming out growth, manufacturing, and drying defects. This approach had distinct advantages since it allowed mills to manufacture a wide range of specialty items of specific grades, including flooring, paneling, moulding, and partition. For example, such a mill might accept orders for 1 × 8 C&Btr paneling, pattern 139. With pregraded rough lumber, the finishing superintendent simply selects the required footage of 1 × 8 C&Btr rough stock and dresses it to pattern. Without this capability a mill either runs a pattern from mill run and markets the downfall, or presorts before dressing. Due to the intense labor and equipment requirements this practice is now uncommon in the South.

I once visited a mill that had no sorting capability between the sawmill and kilns. They mixed all widths, lengths, *and* thicknesses for drying. Grade losses were beyond belief.

Figure 2.1. So you think you have problems? How would you like to dry 4/4, 5/4, 6/4, and 7/4 all in the same unit?

Figure 2.2. Double stacking 4/4 for drying. Although cupping is minor in 4-inch widths, it will cause problems in the wider widths.

How Should We Sort?

Here's the best way to sort: sort to solid, uniform unit size, separating items by length, width, and thickness. This results in maximum kiln capacity, highest grade recovery, and eases handling through the system.

Sorting for thickness is essential. Since drying time increases with thickness this will always be an important component of quality drying. High-temperature drying of 4/4 takes anywhere from 8 to 16 hours. For 7/4 this range is 18 to 24 hours. The thinner lumber will always be overdry when two thicknesses are mixed.

Occasionally I find a mill double stacking 4/4 by placing two boards on top of each other within each layer. Complete units are then dried along with a 7/4 charge. On the surface this looks like an acceptable practice, because it saves sticks and the lumber takes about the same time to dry. Mills that are unable to cut sufficient 4/4 for a complete charge within a reasonable time may have no alternative. Keep in mind that, when dried, one side of each board is drier than the other and warps (cups) after dressing. Cupping is not severe in 4-inch-wide stock and is often acceptable in

Figure 2.3. Mixing widths in the same kiln charge. Mix no more than two adjacent widths (4-inch and 6-inch shown) in a charge and place the wider width on top if loading by forklift.

the common grades. Six-inch and wider lumber cups severely after dressing when double stacked for drying. Reduce this cupping by giving rough dry stock sufficient time to equalize in the sheds before dressing, or equalize it within the kiln. Never dry lumber intended for finish or furniture grades in this way.

At a few mills reliable thickness sorting is a problem, or they may not saw enough 4/4 to justify separation. These mills often double stack 4/4 within bundles of 7/4. Separation is done at the planing mill. Two 4/4 boards do not equal one 7/4 in thickness. In addition to cupping problems expect excessive stick loss and degraded 7/4 lumber adjacent to the double-stacked boards. If your mill cuts a full 2 inches (yes, a few still do) stick losses and degrade are less concern, though warp remains. Avoid mixing thicknesses in a kiln charge.

Separate widths for drying and run charges of one width. Computerized analysis of drying rates of different widths show a 3 percent moisture variation when drying 2 × 4s and 2 × 10s together.[1] Separate them if you can. If you must mix widths within a charge, dry widths that differ by no more than 2 inches (6 inch and 8 inch, 8 inch and 10 inch, 10 inch and 12 inch). Dry 4-inch-wide material by itself. Kilns that are multizoned and computer controlled are often able to mix more widths within the same charge. If you set up kilns by forklift and dry several widths in the same

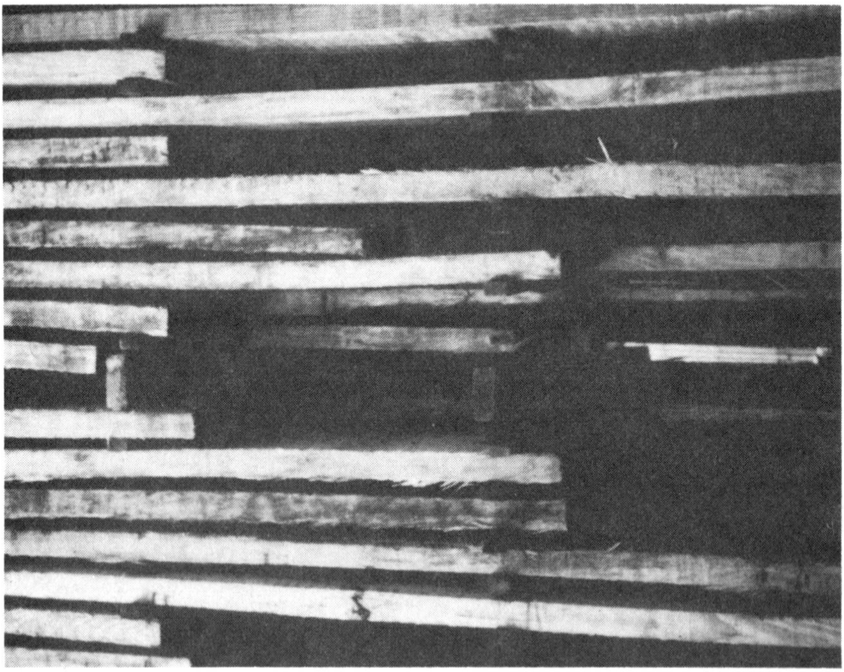

Figure 2.4. Support and degrade problems resulting from mixing too many lengths in a unit. Mix no more than two lengths in a unit.

charge, place the wider widths on top to take advantage of the increased airflow at that position.

Sort all lumber by length. Sorting each separate length is often a problem for a mill. Some were built with insufficient sort capability. More often, new product sorts were added to existing systems, increasing the need for sort bays. In this situation a mill often sorts two or more lengths into a bin. Never sort more than two lengths (8s and 10s, 10s and 12s, and so on) into one bin. If less than half the sorted lumber in a bin is the longer length, consider separating them or reassigning length combinations. Otherwise, expect high trim and stick losses in the overhanging ends. When stacking this lumber always place the longest lengths on the outside of units. This helps maintain even airflow within the kiln and improves lumber support to reduce warp.

Sorting to individual lengths increases average kiln capacity and is a modest means to improve kiln throughput. Add this to any justification to increase sort capacity. Lumber sorted to length is more uniform in final moisture content. Units with mixed lengths have less lumber across the

Figure 2.5. If there are too few boards to insert sticks, unrestrained boards move at will as they overdry. This condition increases trim loss at the planing mill.

width on one end. This end dries faster since the same amount of heat is spread across fewer boards. It overdries, warping, breaking sticks, and causing degrade. Often the units have so few long lengths that end sticks

Figure 2.6. Mix no more than two lengths in any unit and always insert sticks.

are left out altogether. These ends dry without restraint, moving at will. Trimming of split, warped, and shattered ends becomes accepted practice. When the monthly sawmill and planer production reports come across the manager's desk, does he ask, "where did all the long lengths go?"

Make sure enough boards are sorted into each sort bay to make complete units. In forklift operations one bin load usually fills a single stacked unit. When there are too many boards they are often so few in number that they are dropped out at the stacker. However, a slack stacker crew may simply create an extra layer, raising havoc when the unit is set up to load a kiln charge. Unnoticed, top layers rake off as trucks are pushed in. If there are too few boards to fill the top layer, space them evenly across the top. This distributes the weight of units placed on top, reducing compression damage and saving crossouts and sticks from breakage.

Things get a little trickier in transfer operations. Too many boards often amount to several layers. Deadpile them on crossouts and remove them with a forklift before loading the kiln. Wide variations in stacked truck heights are not acceptable. An extra layer here, a missing layer or two there, and gaps develop under overhead baffles. These openings

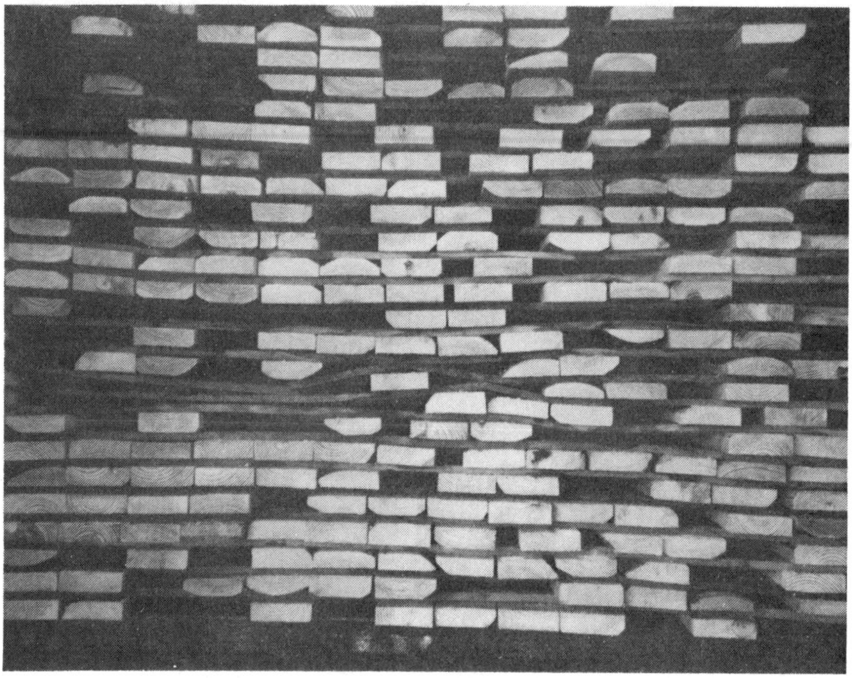

Figure 2.7. Stick distortion leading to eventual breakage. There is not enough lumber to adequately support the load at this end.

allow air to escape around the lumber, reducing overall air velocities within the kiln. Avoid more than two layer differences in height.

Never stack a unit unless you have sufficient lumber to complete the package. Starting a truck of 16-foot only to finish it out with 14-foot leaves a gaping hole for air loss. This has a negative impact on drying.

There is increasing interest and discussion of the merits of sorting for moisture content or wood density before drying. Just imagine how uniform final moisture distributions would be if everything going into the kiln were the same. We could shorten drying time, reduce degrade, and not worry about moisture kickbacks from the planing mill. On a practical basis this idea has not progressed too far in the South. While it is possible to make the separations (the Canadians have developed an infrared moisture detector for green lumber reportedly accurate to 200 percent moisture content), we increase the need for sort capacity. Sort two moistures and we double the required sort bays. Sorting three moistures triples it, and so on. The capital cost of sorting equipment and the extra handling it

Figure 2.8. Too much lumber in a stacked unit that has been deadpiled on top for later forklift removal.

requires will limit its initial acceptance to mills with limited products. A stud mill drying mixed peeler core and green timber production is a prime candidate to test this technology in the South.

Green Lumber Handling

There are two categories of green handling methods. One is found at mills that sort units for forklift handling. The other sorts for transfer cars.

Forklift unit size is limited by forklift weight capacities, and sorter bins are designed to hold one complete unit. The number of units required per kiln cart also depends on kiln door height and the maximum lumber length sawn. A mill might stack two units for 12-foot and under with three for everything else. Five units may be required for tall doors and 20-foot and longer lumber. I recommend keeping the number of units per cart constant for uniform airflow. Use the fewest units possible for your operation, since each unit is separated by crossouts. A lot of air escapes through these crossouts. Many people advocate transfer operations because they lack crossouts. Budgeting forklift replacement this year? Consider higher capacity forklifts to reduce crossouts, keeping in mind sorter bin capacity.

Figure 2.9. Oops! Ran out of lumber. This gap reduces airflows in its vicinity. Some folks try to close this gap by nailing up plywood baffles.

A great deal of green end handling equipment is too light for green end use. If the equipment is a chronic maintenance problem, investigate replacing it. As you compare the equipment from several suppliers, compare the total weight of steel in each. Look at the horsepower. Low initial purchase prices are soon offset by downtime and repair costs. Go for rugged, overdesigned equipment when considering sorting and green handling equipment. You will not regret the investment.

Figure 2.10. Lumber bundled and held in windrows for stacking. Avoid piling these units up on top of one another.

Storing Green Lumber in Windrows

Lumber is often simply scooped up loose at the sorter by forklift and transported to long windrows for temporary storage. These windrows are organized by laying down wooden or concrete sills designed to keep lumber off the ground. More often it does not. Individual sort bundles may be banded or chained into a unit. This increases handling costs but reduces downfall between the sawmill and kilns. Even here, loose and broken boards pile up and are lost from the system. Ends lay on pavement or in mud, picking up rocks and grit that reduce planer knife life. These operations often have an unkempt appearance and high downfall factors. They also have more planer downtime.

To reduce losses from running a windrow storage system, pave the storage area with concrete or asphalt. Simple graded and graveled yards are not acceptable. These yards rut and are a ready source of rocks and grit for your planer. Provide adequate drainage. Keep the area clean, sweeping occasionally and removing trash and broken boards. Align forklift alleys north/south to keep them dry in winter. Position two sills for each row. One sill is not adequate since one end seems to always rest on the

Figure 2.11. Windrow-held lumber with one end resting on the ground. The ends pick up rocks and grit, which hastens knife wear at the planer.

Figure 2.12. Bundle units at both ends and place on two supports to keep the lumber off the ground.

Figure 2.13. A well-organized green yard. Lumber is accessible from both ends of the row to keep lumber turned around.

ground. Space sills one-half the length of the lumber planned for that row. For example, rest 16-foot lumber on sills 8 feet apart. Give yourself plenty of walking room (3 to 4 feet) around the lumber ends for inventory and inspection. You will appreciate it later. Anchor sills by pinning them to the ground with rebar rods. The sills may be wood, concrete, or steel, 6 inches thick and at least 6 inches wide–the wider the better. At the sorter place two bands or chains around each sorted bundle. Never move lumber loose or with one band. This prevents lumber within the bundles from crossing up and breaking. Store bundles one deep to prevent breakage. Make each row accessible from each end to allow the lift driver to place fresh green lumber on one end of the row and stack older lumber off the other side. Otherwise the last bundle in a row always seems to hang around too long. Stacking and placing lumber into the kiln as soon as possible after manufacture gives us highest lumber quality.

Transfer operations stack entire kiln carts at the stacker. Lumber sorted for these operations is held on a green yard until there is sufficient lumber to stack a cart. If lumber is held in windrows, setting up the green yard follows the same rules as a forklift yard.

Storing Green Lumber in Deadpiled Units

Many mills run two stackers, with the second one at the sorter. The contents of a sort bin are stacked into units without sticks (deadpiled) and

Figure 2.14. Green lumber held on sticks. Note the bottom layer distortion that results from insufficient ground support and misplaced crossouts. This unit needs at least three supports, with the crossouts located in the stick row above them.

Figure 2.15. Mixing green lumber with lumber that is seriously blued invites moisture complaints from the planing mill.

banded before transfer to the green yard. This method saves yard space and reduces degrade when compared to windrow storage, but requires more labor. The green storage yard is set up the same as for stickered lumber in a forklift operation (see page 32). Want to save the labor and expense of banding? Try this method in use at several mills: As successive layers are deposited by the stacker, a small air cylinder automatically pushes every other layer over an extra few inches on the even end. Since boards do not line up one above the other in the package, the unit "knits" together and can be moved about without bands.

The most efficient forklift operations operate with the lumber stacker direct coupled to the sorter. This forces a higher level of maintenance on the sorter and stacker, since unscheduled downtime eventually shuts down the sawmill. When people are expected to operate at a higher level they develop the attitudes necessary to get the job done. These attitudes carry over throughout the mill.

Direct coupling is the least expensive alternative. By forcing the mill to maintain its equipment at a higher level, the entire sawmill runs better. The extra handling costs required to move green lumber in and out of storage prior to stacking are eliminated. Grade and volume loss is held to a minimum. All lumber is stacked and held on sticks to get a leg up on drying.

Figure 2.16. Even subtle differences in lumber dryness as it enters the kiln show up after drying.

Figure 2.17. Good sorting and handling improves your chances of avoiding drying problems.

Storing Green Lumber on Sticks

Still, this stacked lumber must be held in storage before placing in the kiln. Grade and pave this storage area to give the forklift a level working surface. Rough, rutted roads cause boards to fall out and sticks to shift, degrading lumber. Provide good drainage. Align alleys north/south to keep them dry. Stickered lumber is flexible and requires good support. Oak sills, 6 × 8 inches and larger, make good foundations when arranged to line up with stickers. Place two sills under lengths to 10 feet, three under lengths to 16 feet, and four under everything else. Add additional sills if you see bottom layers sagging. To reduce crossout handling, use crossouts from the kiln. I recommend 4 × 4s located at each stick location. Some mills use 2 × 4s for crossouts, but I find they shift easily and fall over. They do not last as long, either. A few mills have found that they can use 2 × 4s flat in conjunction with thin forks on the forklift.

Always keep your lumber as even-aged as possible by stacking the oldest stock first. This avoids an important cause of spotty wet lumber. One week is the limit for holding lumber before stacking. Fresh lumber mixed with lumber over a week old is difficult to dry to uniform moisture content.

Most transfer operations move entire loaded kiln carts from the stacker through the kilns to the planer breakdown. Some insert crossouts for forklift handling behind the kiln. These mills lose the primary benefit of transfer operations, elimination of the crossout.

Compared to a well-run forklift operation, transfer operations are labor and equipment intensive. Nevertheless they seem to be favored by the larger mills. Transfer cars endure much abuse and breakdowns are frequent. Cables used to move cars and pull kilns often break or slip off, causing injury. I once watched amazed as a cable suddenly broke and caught a man's ankle, turning him head over heels. He was fortunate—only the cable was broken. Cart wrecks, in and out of kilns, are more frequent. Kiln trucks also sustain more damage and incur higher repair costs.

Whichever system you use, place all lumber into the kiln as soon as possible to reduce surface drying and fungal degrade. Fresh green lumber is easy to dry in a high-temperature kiln. Lumber held on the yard long enough to blue stain takes a little longer to dry. However, lumber that is blued and a month or more older takes less time. What is important is that each kiln charge be composed of lumber of a similar age. This narrows final moisture distributions and improves grade.

CHAPTER GUIDELINES FOR REDUCING VARIABILITY IN DRYING

Sorting

1. Sort and dry lumber by thickness. Never mix thicknesses within the same kiln charge.
2. Sort and dry lumber by width. If you must mix widths within a kiln, they should vary by no more than 2 inches.
3. Sort lumber by length. If you must mix lengths, mix no more than two lengths in a unit and place longer lengths on the outside.
4. Sort enough boards into a sort bin to fill complete units. Remove excess boards before kilns. If the top layer is not filled, space the boards evenly across the width to support the unit above it.
5. Never mix package lengths on a kiln truck, this creates large gaps for airloss.
6. Make bundles uniform in height and width.

Green Lumber Handling

Forklift Operations

1. Use the fewest number of units possible per kiln truck to reduce crossouts. Keep the number of crossouts constant between trucks.
2. Run the stacker direct coupled to the sorter. Avoid handling unstacked lumber.
3. If you must yard green lumber before stacking, deadpile into units or band into bundles. Use two bands or chains per unit.
4. Remove excess boards from the system; do not add extra layers. Space boards in unfilled layers across the package width.
5. When storing stacked green lumber prior to placing in kiln, place a crossout at each stick location. Place lumber on sufficient level and aligned sills to prevent the bottom layers from sagging.

Transfer Operations

1. When yarding green lumber for stacking, deadpile into units or band into bundles. Use two bands or chains per unit.
2. Completed stacks should be the same height. Avoid variations in stack height that exceed 2 layers.

For Both Systems

1. Stack and place lumber into the kilns as soon as possible after manufacture. Lumber in any kiln charge must be as even aged as possible to reduce drying problems.
2. Plan your loading of kiln charges rather than letting it happen!

Guidelines for Establishing a Green Handling Yard

1. Grade and pave area for a level and dry surface. Provide good drainage. Keep area clean and free of debris. Align alleys north/south so the sun can dry the pavement and melt snow.
2. Position two support sills for each row, spacing them one-half the length of the lumber they will support. Anchor them.
3. Sills may be rot-resistant wood, concrete, or steel, 6 × 8 inches or larger.
4. Space adjacent rows of lumber with 3 to 4 feet between for inventory access.
5. Make each row accessible from each end to keep stock rotated on a first-in, first-out basis.

6. Band each bundle with two bands, one on each end about 2 feet back from the end. Do not store more than 1 bundle deep.

Guidelines for Establishing a Stacked Green Lumber Yard

1. Grade and pave the yard, providing good drainage. Align alleys north/south. Keep the yard clean.
2. Use sills made from rot-resistant wood, concrete, or steel, spacing them to line up with sticks and eliminate sagging.
3. Place crossouts at each stick location or at least at the stick where the sill is positioned.

REFERENCES

1. Denig, Joseph, Stephen J. Hanover, and C. Arthur Hart. *Kiln Drying Southern Yellow Pine Lumber.* The North Carolina Agricultural Extension Service. Publication No. AG-402, 1988.

CHAPTER 3

Lumber Stacking and Sticklaying

If a man is called to be a streetsweeper, he should sweep streets even as Michelangelo painted, or Beethoven composed music, or Shakespeare wrote poetry. He should sweep streets so well that all the hosts in heaven and earth will pause to say, here lived a great streetsweeper who did his job well.

Martin Luther King, Jr.

Improper stacking and sticklaying is often the greatest source of degrade in a drying operation. To understand why, we must visualize what happens to lumber during the drying process.

High-temperature dried lumber goes through a "plastic" stage during the high heat and humidity of warmup. The conditions are similar to a method cabinetmakers use to bend special wooden shapes for chairs and furniture parts. For example, here's how a craftsman forms the curves from straight lumber to make a bentwood rocker: To form the rocker's arm the craftsman selects dry, straight-grained lumber and machines it to the desired cross-sectional shape. He then places it into a steaming chamber for about an hour, until the wood becomes pliable. After it is removed from the hot, moist environment of the chamber he bends and clamps the wood into the form, which determines its shape. Once the arm is cooled, he releases it from the form and works it into the chair. The arm will keep its shape.

This same process occurs in the high-temperature kiln on a grand scale. Misplaced or missing stickers and those that vary in size do not hold lumber flat. Wherever there are variations in support the lumber bends to conform to the forces placed on it. Once dried, the distortions are permanent. This shows up in dressed lumber as short skipped areas on the wide face.

Correction of poor stacking methods is easy, but it is a continuing effort. Don't let up. The rewards are there. Good stacking becomes natural and automatic as personnel get into the habit of doing it right. When training new people, explain *why* they are stacking lumber as well as

Figure 3.1. Poor stick quality and placement degrades lumber.

demonstrating how it is done. Take them to the dry sheds behind the kilns and show them the impact of poor sticklaying. Most people want to do a good job, and will—if they understand that what they are doing is important.

Stick Material

An amazing variety of stick materials is available. What is most economical for your operation depends not only on initial purchase price but also on the factors that affect stick life. Do wide variations in lumber thicknesses distort and break sticks? Do you handle stacked lumber by transfer or forklift? Does stick handling and storage equipment cause breakage? All these factors and more affect the actual cost of sticks. Determine which stick is best for you by testing different ones.

Estimate actual stick costs by dividing the cost of a stick by the average number of times it passes through the kilns. This is best determined by color coding the ends of a new unit of sticks and tracking them through the system. After each pass through the kiln, count them. When half remain,

Figure 3.2. Short skipped areas on dressed lumber caused by misplaced sticks.

you know your average stick life in charges per stick. Looked at in this manner, an inexpensive stick with a short life costs more than you think. While conducting this test, try to determine why your sticks break. Is it their quality or are you doing something that shortens their life? Mismanufactured lumber, mixed lengths within stacked units, poor stick handling, and other factors all affect stick life. Each can be corrected.

The following sticks are available:

Self-produced or Purchased Domestic Hardwood Hardwood sticks are highly variable in quality unless kiln dried to 10 percent moisture content before sizing. All that I have seen were green when first used to stack lumber. Wood shrinks when dried, and shrinks at different rates depending on such factors as species, green moisture content, and the orientation of

Figure 3.3. Poor stacking of flitches. This unit is in danger of toppling over.

growth rings. Add variation in sawn green dimension and the final thickness of dry sticks in service is remarkable. This variation distorts the lumber it supports. The actual service life of hardwood sticks depends on the grade of wood they are made from. Grain variation around knots

Figure 3.4. Hardwood sticks held in the original units from the supplier until they mold and degrade. Place green hardwood sticks immediately into the kiln to prevent degrade.

shortens their life. Tapered wane causes warp. Clear, straight-grained, dry hardwood sticks are prohibitive in price due to their value as clear lumber. Still, many mills continue to use green hardwood sticks, perhaps lulled by low initial purchase price. The actual cost of these sticks is very high due to poor lumber support and short stick life.

If you insist on using green hardwood sticks be very careful. Inspect each shipment for uniformity and turn back loads with poor thickness control and excessive wane. Place them immediately into the system to get their moisture content down. Remember, they are green and subject to mold and decay. I have seen mills hold these sticks for months before placing them into the system. By then, decay shortened their life to one charge.

Self-produced or Purchased Pine Pine sticks are often produced at the mill from dry stock specially graded out at the planer. Graders select clear 2 × 8s and wider that have Economy or saddle wane. Sticks are cut using cutoff and gang rip saws. High-quality #1 dimension lumber is good if the need arises and the market is slow. All of these sticks are straight grained and have few knots to limit their life expectancy. They make good sticks, although you may notice a tendency for them to become thinner with use. Never make sticks from #2 common and lower grades—the knots and cross grain cause too much breakage. When you consider the initial material and manufacturing costs, these sticks can be pretty expensive.

Phenolic Resin or Plastic Sticks Occasionally you will find phenolic resin or plastic sticks in the South, although their use is rare. I know of no one who uses them on a large-scale basis. Some concerns include the use of petroleum-based products for their manufacture; the disposal or recycling of broken sticks; and the tendency for plastics to become brittle under continuous high temperatures. One stick company buys back broken sticks at about 25 percent of the purchase price and recycles them.

Aluminum Sticks At least one mill in the South has made trial runs using extruded aluminum sticks. These sticks represent a significant investment that would be economical were it not for their tendency to bend rather than flex under load. Once bent, their life expectancy is very short. The primary concern governing their use is the recycling of bent and broken sticks. They are also hot to handle if not allowed to cool between uses; and over time they are reported to develop sharp burrs, which cut sticklayers' hands.

Tubular Steel Sticks This interesting variation on the aluminum stick seems to have caught on at a few mills. These sticks also bend under load, but seem to be a little easier to use. Broken sticks are welded together into longer lengths and cut to normal stick length. Their initial expense and the repair costs necessary to keep them in service reduce their competitiveness. Also, after a few repairs, they are not straight and do not support lumber evenly.

Plywood Sticks Plywood sticks are becoming more common as plywood plants seek to diversify their product mix. In general, these sticks are very good. There are two main types. The first has its layers laid flat across the width of the stick. They give good service in high-temperature kilns. A newer plywood stick has its layers laid flat on the narrow edge, showing glue lines on the wide face. This stick is stronger and gives longer service life. All are uniform in size and dry when purchased. They also tend to keep their dimensions better and warp less than solid wood sticks. I have heard complaints of delamination or separation of layers in these sticks. This is easily corrected by the manufacturer.

Apitong Sticks These imported hardwood sticks are good for mills with high stick losses resulting from poor lumber size control. They are highly durable and last a long time under severe service conditions. They are uniform in size and keep their shape well. Western mills have used them for years with good results. Although they cost more than domestic sticks, they last two to three times longer. Their main drawback is the length of time it takes to get them once you place an order—usually two to three months. One note: Apitong is a high-density hardwood, which makes these sticks a bit heavier than most. You may hear some grumbling among sticklayers who are used to handling lighter sticks.

OSB Sticks (Oriented Strand Board Sticks) On a recent trip I was introduced to these new sticks, which were purchased for a trial. These sticks were uniform in size and flexible. As of press time there were no performance data available other than that supplied by the manufacturer.

Sticker Size

Thickness Thickness is the most important dimension in a stick, since it determines the spacing for each layer. Thicknesses must be uniform to provide even support to lumber without degrade. There must be free flow of air, since that is the reason for stickering in the first place. Mills that produce their own sticks must monitor their manufacture to assure quality. Incoming purchased sticks must be spot checked and rejected if poorly made or the wrong size. Quality is often variable.

Use the thinnest stick that gives good service life and the best drying quality. For every $1/8$-inch reduction in stick thickness you will gain an average of 5 percent in kiln capacity. Be careful here: Increases in kiln capacity are easily offset by high stick losses, increased lumber degrade, extended drying time, and wider moisture distributions.

Most mills use $3/4$-inch or $7/8$-inch sticks and enjoy long stick life—provided they have good lumber size control. Studies show a negligible impact on drying time between these thicknesses, although I prefer $7/8$-inch sticks.[1] It is interesting to note that as you reduce stick thicknesses within some kilns, air velocities increase. However, since thinner sticks increase total lumber volume, any benefit to drying time is often offset by the increased heat demand. If your fan system is already performing near its maximum static air pressure capacity, air velocity may not increase.

A few mills use $1\,1/8$-inch to $1\,1/4$-inch sticks. They pay a steep price in lost kiln capacity. Many of these mills do this to extend stick life and improve airflow due to poor lumber size control. Others have kilns with marginal heat capacity and seek to improve drying speed and lumber quality. Mills that dry a large proportion of timbers benefit from the extended life of thicker sticks. If your fan system is undersized, thicker sticks tend to give more uniform drying across the load.

Studies done at Oregon State indicate that stick thicknesses as small as $^{7}/_{16}$ inch are practical for conventional kilns.[2] At least one mill study has been done comparing $^{5}/_{8}$-inch sticks to $^{3}/_{4}$-inch sticks in a high-temperature kiln. The $^{5}/_{8}$-inch sticks do not seem to perform as well. My primary concern about the thinner sticks is their service life. They are intolerant of variations in lumber thickness.

Width Sticks must be wide enough to assure that sticklayers recognize the difference between width and thickness at normal stacking rates. One-and-a-half inches is a common width. Square sticks are acceptable in $1^{1}/_{8}$-inch to $1^{1}/_{4}$-inch sizes, but avoid smaller ones due to reduced stick life. Some sticklaying machines require sticks $2^{1}/_{2}$ inches to $3^{1}/_{2}$ inches wide to work right. This may be good for the machine, but it is bad for drying. Expect wet spots under the stickers. Do not buy a sticklaying machine that requires these sticks to function reliably.

Length Sticks must be long enough to extend completely across the stacked unit with a little extra. Never accept sticks that are much shorter than the unit width. Unsupported boards on the edges drop down, covering the airspace, or simply warp in place. If you get short sticks by mistake and must live with them, push the first stick to the far side, pull the second one to the near, push the third to the far, and so on. Although the boards are not fully supported, at least they will not fall and block air circulation.

Uniformity Manufacture all sticks to uniform dimensions for even support. This is often a problem with purchased hardwood or pine sticks. Be cautious and inspect each truckload. Reject stick bundles that vary in thickness, or are waney or knotty. Sticks are expensive enough without the added cost of breakage before use, or with variations that degrade your lumber. Avoid mixing different types of sticks in the system for this reason.

Stack Width

In terms of drying speed and narrowed moisture distributions across the width of a unit, the maximum stack width is around 8 feet. Stack widths of 6 and 7 feet are encountered, which helps narrow moisture distributions still further. However, this reduces the amount of lumber you can place in the kiln. If you are stacking narrow widths and need additional kiln capacity consider going to 8 foot units. Keep in mind that you may have to beef up the heat and/or airflow systems to keep from extending drying time. Still, this will probably be much less expensive than building a new kiln.

I have seen a few kilns designed to handle widths to 9½ to 10 feet. Most were built under west coast influence. Avoid this in high temperature south-

Figure 3.5. Short sticks causing the edge boards to fall down and close sticker openings.

ern pine kilns. Unless these kilns have sufficient air velocity to overcome the large temperature drops as air passes through the lumber, moisture variation across the unit is large and drying times extended. This tendency can be overcome to some degree by using thicker sticks but this reduces kiln capacity, the purpose of the wider widths in the first place.

Mills making the transition from airdrying to kiln drying often keep

Figure 3.6. Short-term fix for the problem in Figure 3.5. Alternate pushing and pulling sticks to the side to support lumber along its length.

their narrow stack widths of 4 to 5 feet and set two bundles side by side on a truck. This is acceptable as a transition, but begin making plans to convert to the wider single unit. When two bundles are placed side by side on a single truck, stick courses do not always line up. If the stacks are too close, wet spots appear in your lumber. Space units 2 to 3 inches apart to allow airflow to reorient from one stack to the next. This stacking

Figure 3.7. Typical problems encountered with sticks. From left to right; crooked stick, broken stick, tapered stick, too thick stick, good stick, thin stick, thick stick.

method also results in twice as many units, which increases handling costs. The additional handling in turn introduces more opportunities for lumber damage and degrade.

Sticklaying

I cannot overemphasize the importance of proper sticklaying. The full cost of untrained or inattentive sticklayers is far more than wages, since it extends to the reduced quality of your lumber. Take the time to show new

Figure 3.8. A mill with no organized plan for removing bad sticks at the stacker. This is a fire and safety hazard.

sticklayers why their job is important; it is not enough to simply tell them. Take them to the dry sheds behind the kilns and point out the distortions caused by poor sticklaying. Most people take a different outlook toward their efforts when they realize how poor workmanship on their part affects the mill.

Stick Quality Maintaining stick quality is a continuing process. Repeated cycles through the kiln and handling equipment cause sticks to warp and break, eventually damaging the lumber they support. Remove any stick with the potential to cause degrade from the system.

Most mills remove sticks at the stacker, relying on the sticklayers to remove bad sticks. High stacking rates make this very difficult since there is little time between cycles. If you depend on your sticklayers to cull sticks, provide a method for stick disposal. Many mills simply position dumpsters near the stacker. Others convey sticks to a nearby hog. The worst practice is throwing them out on the ground for later disposal. This invites fire and creates hazardous working conditions. The best system I have seen removes stickers at the planing mill. As sticks pass by, a man inspects and discards bad sticks into a bin. He may drop them onto a conveyor to the trimsaw hog. Some mills cut them into shorter lengths to use as spacers in finished lumber units. This is a very practical option for operations that bundle stickers going back to the stacker.

Figure 3.9. Sorting sticks at the planing mill. Bad sticks are thrown onto belt on right and conveyed to trimmer hog.

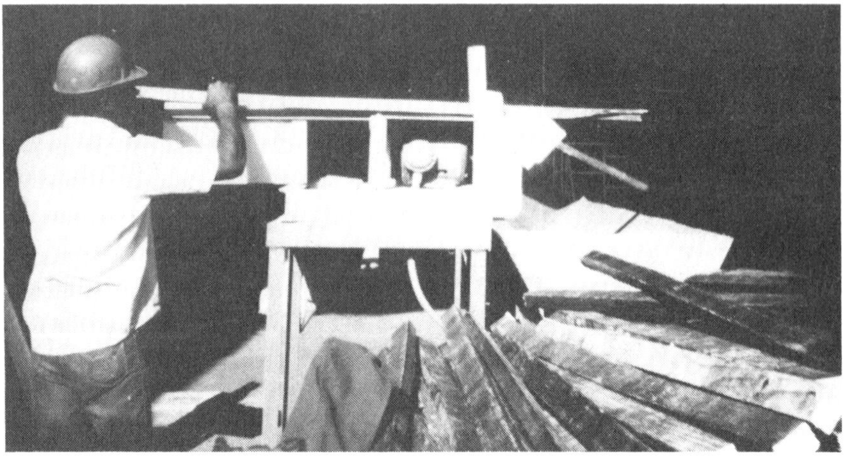

Figure 3.10. Culling and trimming bad sticks at the unstacker to reuse them as spacers in dressed lumber units.

Sticks that are uniform in thickness do not cause degrade. Try to keep stick thicknesses within 1/8 inch of each other. More variation than this results in small distortions that ruin the flat appearance of your lumber. This is important to your customer. Extreme variations degrade lumber and cost you yield. Remove all thick and thin sticks from the system.

Discard broken sticks that do reach across the full width of your units.

Figure 3.11. Piecing short sticks together to extend across the package width. Consider this a temporary measure only since alignment is critical and piecing slows down the stacker.

These sticks cost more in one trip through the kiln than their replacement cost. Unsupported edge boards warp at will. Boards drop down and cover the sticker space. The reduced airflows widen moisture distributions. If you are in a bind for sticks, lay two sticks side by side to extend across the width. Consider this a temporary measure only, as improper placement impedes airflow.

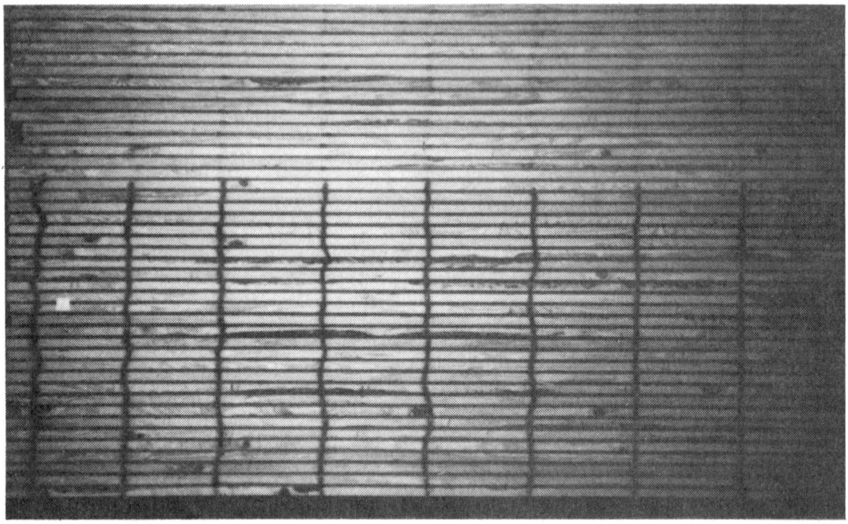

Figure 3.12. Good stick alignment on the side away from the sticklayer.

Crooked sticks upset airflows and produce wet and overdry spots in your lumber. On the outside of a curved stick, air speeds up due to the restriction and drying is faster. The inside of the curve creates a dead air space, which slows drying. The resulting moisture variations down the length of a board cause quality problems for customers using your lumber for furniture or laminated beams.

Stick Placement Lay straight, uniform sticks perpendicular to the length of the unit, spaced on 2-foot centers. Some mills stack on 4-foot centers or stack only the bottom 10 or so courses on 2-foot centers, the balance on 4-foot centers. They claim that their studies indicate no additional degrade from this practice. One thing is certain: Stick alignment becomes more critical with fewer sticks. I tend toward the conservative in this area and recommend 2-foot centers for sticks and truck supports. Always use stick guides that align with your kiln truck supports. Make a scale drawing to determine accurate stick positions. This also allows you to determine spacing for other equipment that is affected by stick spacing.

Good stick guides are sometimes hard to keep in service. The best guides form pockets on both sides of the load into which the sticklayer drops his sticks. For many stackers this is mechanically impossible on the far side. Alternatives include hanging short, heavy chains from a rugged support to mark stick locations. Lasers and shadowlines also work well, but be careful with shadowlines. The high temperatures older ones produce may cause a fire if the area is not kept clean.

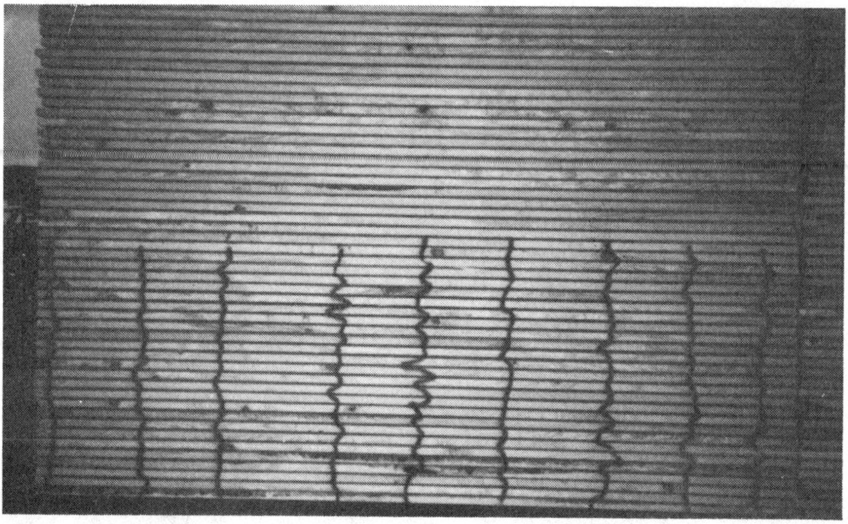

Figure 3.13. The side next to the sticklayer illustrates their tendency to simply drop the end they are holding, rather than place it.

Mechanical sticklayers are becoming common. Be very careful when considering their purchase. Look for rugged but simple equipment. Equipment with a lot of small working parts to maintain are difficult to keep in peak performance operation. The best ones I have seen position each stick rather than drop them from a hopper. All require high stick quality, but some have problems laying sticks consistently right. The most common problems I see are missing sticks and sticks laid on edge. Both cause degrade. Evaluate the available equipment very carefully to make certain the unit you buy does not cost you down the road. Visit mills running the equipment and watch it work. Walk through the stacked green lumber yard and look for missing sticks and those turned on edge or misplaced. Finally, ask the manager if he noticed any grade improvement after installing the mechanical sticklayer. A good one actually improves grade.

Lay sticks one above the other for continuous support from top to bottom in the unit. Unaligned sticks kink the wood, which shows up as small localized skips at the planing mill. Check stick alignment by spray painting a line connecting each stick top to bottom in a unit. Properly stacked units have straight lines. This method helps identify sticklayers who need improvement. Paint the sides of a stacked unit and place it in clear view of the sticklayers. You need not say anything to them. Sooner or later one of them will come to you wondering what you are doing.

Figure 3.14. Painting the ends of sticks is a good method for evaluating sticklayer performance. The sticklayer on the left needs coaching.

Reply that you are checking stick alignment and demonstrate the impact of poor stacking practices. The improvement will surprise you.

This is an effective way to deal with sticklayers who are slow to learn. A sticklayer who is glaringly compared to his workmates would be stubborn indeed to ignore the need to improve. Peer pressure is a powerful force. Still, there are always those who refuse to change. Photo documentation over a reasonable period of time should be grounds for dismissal or reassignment. You cannot afford to have these people on the job. A final note: "Connect the sticks" on both sides of the unit and compare both sides. Sticklayers often position their sticks in the far pocket and simply drop the end they hold. Look for wavy stick alignment on their side.

Avoid turning sticks up on edge. This causes serious distortions in the lumber. If you notice this a lot it may indicate that you are running the stacker too fast. Rushed sticklayers cannot pay close attention to detail. Sometimes sticklayers intentionally turn sticks up on edge or double stack sticks on one end to keep the lumber surface level. Look for one of two causes. First, are you sawing taper into your lumber? As layers mount up, one end rises faster than the other. Stackers do not operate well unless the lumber is level. Sticklayers compensate for this by doubling up on the low end, usually without comment to the supervisor. Sometimes the stacker

Figure 3.15. Stick accidentally laid on edge. When dried, this layer will have a slight bow in the end.

itself is not level or loses its level as it loads. I have seen this on stackers whose cables were replaced on one end only. The stretch difference between the two sets of cables increased as the load increased.

Ideally, we want a solid lumber layer across the unit width with even edges on the side for best drying. Wide variations in lumber target widths make solid layers with even edges difficult. Full-width layers alternate with others lacking a board. There is variation in total width between full layers. Here we must compromise a bit to achieve our goals. Encourage sticklayers to pull side boards to the outside to form an even edge. This improves airflow uniformity between layers. However, pulling boards to the side leaves a gap in that layer, forming a chimney. There is always a gap in 2 × 10s. These solitary edge boards are drier and prone to warp. Eliminate the chimney and reduce dryness by pulling back one board on the first layer, two on the second, one on the third, and so on as layers are stacked.

Be careful if you allow sticks to protrude on the outside. Mills operating tilt hoists in the planing mill will sustain serious stick losses unless they turn protruding sticks to the outside. Exposed stick ends simply shear off.

Figure 3.16. Haphazard stacking resulting in uneven unit edges. The uneven edges affect airflow uniformity between layers when this side is next to the plenum chambers.

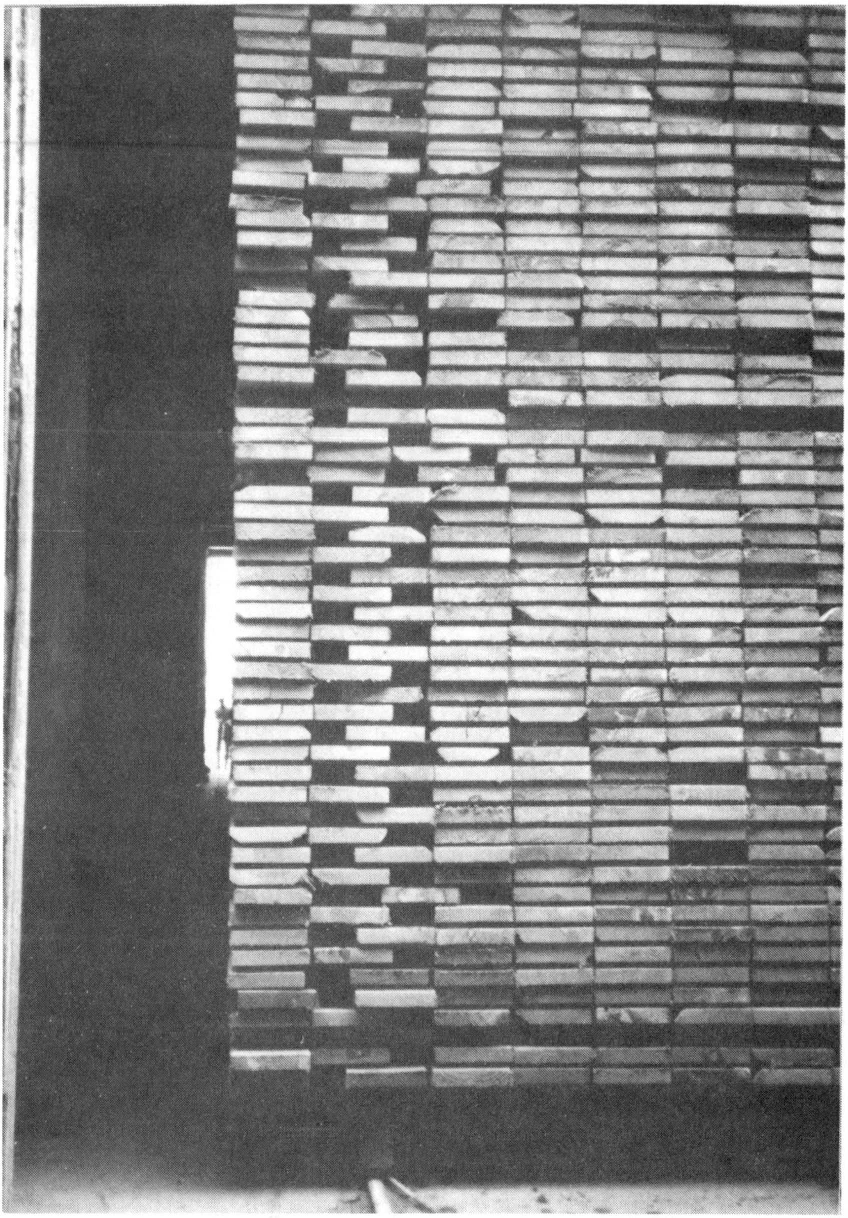

Figure 3.17. Good stacking technique, which helps maintain uniform airflow.

Figure 3.18. An unstacker should be gentle on the sticks and the lumber.

Care and Storage of Sticks

There are two common methods for transporting and storing sticks between the planing mill and stacker. Most transfer operations operate on a closed loop system, conveying sticks directly from the unstacker to the stacker. Many of these systems use a rake-off method of unstacking that is very gentle on the lumber and sticks. Stick breakage is low. They do not sit around on the ground and deteriorate. Introducing new sticks to the system is often difficult unless provisions for this are designed into the original installation.

Most forklift operations use a tilt hoist method of unstacking. They bundle sticks or gather them into racks for transport back to the stacker. This is tough on sticks. Rising hoists shear sticks off on one end. As they slide off onto the conveyor, weakened sticks break. Gathered into racks or strapped into bundles for forklift handling, still more breakage occurs. Storage takes its toll as bundles heap on one another. Sticks represent a significant investment for any mill. You must train personnel to protect that investment. Carefully stack them in racks or bundles. Store bundles off the ground and no more than one high. For long-term storage, place them under dry shelter.

A good friend in North Carolina offers this wise advice to stacker supervisors: "Care enough about your sticklaying to show that you are interested, today and everyday. Only then will you see lasting results."

Figure 3.19. Poor stick storage. Although the container is a good idea, poor loading methods reduce stick life.

Figure 3.20. Proper use of a stick rack.

Figure 3.21. A stick bundling method at the unstacker that reduces stick breakage and allows bundling of sticks into units to return to the stacker.

Figure 3.22. How you handle bundled units of sticks has a significant impact on how long your sticks last. Store sticks off the ground, one bundle high. The new sticks in the background must be placed into the kiln as soon as possible to reduce degrade.

Figure 3.23. Our goal: a perfectly stacked unit. This unit has been stacked for a transfer system.

CHAPTER GUIDELINES FOR REDUCING VARIABILITY IN DRYING

1. Train your sticklayers by showing them the effect poor sticklaying has on lumber.
2. Determine the most economical stick for your operation by comparing service life to initial cost.
3. Use sticks 3/4 inch to 7/8 inch thick by 1 1/2 inches wide. Timbers may require thicker sticks to extend their service life.
4. Sticks must extend full across the width of your units.
5. Remove thick, thin, crooked, and broken sticks from the system.
6. Align sticks one above the other in your units for complete support.
7. Lay sticks on 2-foot centers to align with truck supports.
8. If sticks begin appearing turned on edge or double stacked, find out why and correct the cause.
9. As they stack each layer, sticklayers should pull the edge boards toward them to form a square-edged package.

Figure 3.24. A well-stacked unit for a forklift operation. It is hard to believe that this lumber was stacked by hand off a line sorter. This precision is the envy of many stacker operations. If the forklift driver would place crossouts at each stick location, I could find nothing wrong with this picture.

10. Finally, take care of your sticks between the unstacker and stacker to extend their life.

REFERENCES

1. Denig, Joseph, Stephen J. Hanover, and C. Arthur Hart. *Kiln Drying Southern Yellow Pine Lumber.* The North Carolina Agricultural Extension Service. Publication No. AG-402, 1988.
2. Horton, Robert K., and Helmuth Resch. "Kiln-Stick Thickness Effect on Drying 2-inch Western Hemlock Lumber." *Forest Products Journal* 26, No. 3 (1976): 35–40.

CHAPTER 4

Kiln Setups and Loading Practices

Genius is the ability to reduce the complicated to the simple.
C. W. Ceran

Once lumber has been manufactured, sorted, and stacked it is ready to load into the kiln. How well we set up lumber for loading has a significant impact on how well it dries. Our goal at this point is to dry the most lumber in the shortest time with the highest quality. This maximizes the efficiency of our operation and reduces costs.

Selecting Lumber from Stacked Green Inventory

Lumber going into the kilns must be as even aged as possible for highest quality drying. Fresh-sawn lumber dries at a different rate from lumber that has been on the yard long enough to surface dry. Stained lumber dries differently still. Mixing lumber within the same charge results in wet and overdry units at the planing mill.

Inventory stacked lumber each morning and set up kilns from this list, drying the oldest lumber first. In general, consider as even aged all lumber manufactured within any seven-day period and dry this in the same charge. Lumber over a month old is considered even aged and dried as partially air-dried stock.

Design plenty of drying and boiler capacity into your sawmill system. Mills with marginal or inadequate kiln capacity leave no room for error. A well balanced sawmill and drying operation is able to turn all sawmill production within six days. This leaves one day for routine maintenance and unexpected downtime. Without this cushion you can expect unforeseen problems to cover you up with green lumber.

We have already discussed several points to keep in mind when setting up your kilns. Let's review them here:

Figure 4.1. Never try to dry lumber of different age or condition within the same kiln charge.

1. Never mix thicknesses in the same kiln charge.
2. Mix no more than two widths within the same kiln charge. If you stack for forklifts, place the wider lumber on top. Dry 2 × 4s by themselves if possible.
3. Never mix unit lengths on the same truck.

Keep your stacked green yard as small as possible to maintain control. This is easiest for forklift yards with computerized multizone kilns mixing widths in the same kiln charge. These mills often store green stacked lumber by length alone. If you must separate widths then your yard must be much larger. Follow the guidelines in chapter 2 for setting up these storage yards.

Transfer operations need sufficient storage track for a complete charge of each width. This storage may include the green end kiln tracks when large amounts of a few widths are produced. These widths are loaded directly onto tracks to the next available kiln. Leave enough storage space for those odd units you drop out to get the right length combination for your kiln. Don't forget to put them into the next available kiln charge.

Figure 4.2. Keep unit heights the same and never mix thicknesses in the same kiln charge.

Find a system for tracking lumber from the stacker to the kilns that works for you. The options vary from spray painting item and stacking dates on the sides of units to attaching tags with this data. (Sample tags for this purpose are illustrated in chapter 13.) New, computerized systems read barcoded numbers to inventory lumber and set up kilns for optimum drying efficiency. They are valuable for transfer operations handling large volumes of lumber.

Kiln Trucks

Good-quality kiln trucks with supports under each stick location are essential for high-grade drying. Four-foot-long trucks with three bunk supports on 2-foot centers are standard for southern pine. These trucks are easily combined in different combinations to support all lumber lengths. This assures continuous lumber support and weight transfer to the ground for distortion-free lumber.

Many mills cut corners by placing only two supports on a 4-foot truck. Their initial investment saving is soon wiped out through increased lumber degrade and shortened truck life. Bottom lumber layers distort with a characteristic "kiln kink" typical of these operations. Trucks them-

Figure 4.3. A standard southern pine kiln truck. It is 4 feet long and has three supports spaced 2 feet apart.

Figure 4.4. A 4-foot kiln truck with only two supports. Note the bending in the truck from the lumber load.

selves bend as weighted ends curve down under load. This distortion changes the actual height of the truck and new trucks added to the system appear higher. As multiple truck heights mix together in the system, bottom layer degrade becomes severe. Once distorted in this manner, you must shim each truck individually to add a third support. New trucks will still be higher. The best solution is total replacement of all kiln trucks at one time. If this is not possible due to financial constraints, keep truck styles separate as you proceed with the conversion.

The best bunk supports are made from square tube or from two pieces of channel welded to form a tube. I-beam supports are easily bent in transfer operations and they catch forks in forklift operations. Weld bunk supports permanently in place to keep them in the right position. Welded trucks are easier to move about by forklift. Never use wooden bunk supports. They vary widely in dimension and must be handled and positioned each time they are used. This degrades lumber.

The minimum recommended combinations of 4-foot kiln trucks for each length combination is as follows:

Table 4.1 ***Combinations of 4-foot Kiln Trucks and Lumber Lengths***

Lumber Unit Length	*Minimum # of Trucks for*	
	Forklift Operations	*Transfer Operations*
8′	2	2
10′	2	2
12′	3	3
14′	3	3
16′	3	4
18′	4	4
20′	4	5
22′	4	5
24′	5	6

I prefer to use the transfer system numbers for both operation methods. This assures maximum support with minimum slippage while moving loaded trucks about.

Avoid 6-foot trucks unless you stack 6-foot lumber. These trucks make odd combinations for setting up the various lengths going into the kilns. Purchase only a few to cover your 6-foot production. Be very careful with them, since some have insufficient support for the weight they are asked to carry. Either beef up the side rails or add extra wheels to avoid truck distortion. If you insist on using them make sure supports are on 2-foot

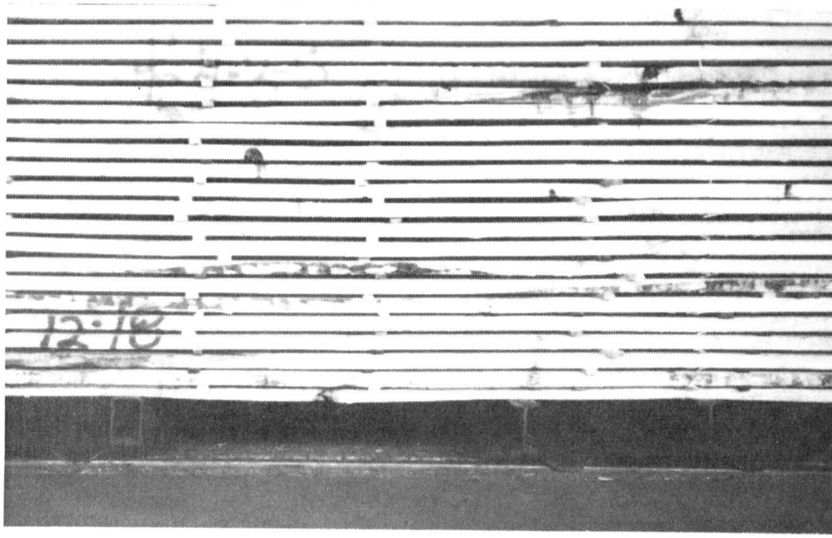

Figure 4.5. Two styles of truck supports. The left side shows two pieces of channel welded together to make a box beam. The right is an I-beam. Note the lumber distortion that results from insufficient and misplaced support.

Figure 4.6. Wooden truck supports that are too few and poorly positioned to support the load.

centers to align with the sticks. I once visited a mill using 6-foot trucks with supports on 3-foot centers. These trucks did not line up with anything.

Use 8-foot trucks only at stud mills.

I have seen two examples of a single, continuous truck for each kiln track. One kiln was 84 feet long. Due to the large number of possible lumber length combinations to fill the truck, supports were unattached. In theory this allowed the lift driver to position each support where needed. In practice, however, few lift drivers will do this consistently on a continuing basis. The grade losses from this truck were staggering. Some mills weld the truck supports in place. But be cautioned that it is difficult to place these supports in a configuration that accounts for all lumber length combinations.

Use wheels fitted with grease fittings on the end of the axle and lubricate them once a week. Some operations grease them each time through

Figure 4.7. An example of a single, full-kiln-length truck with moveable supports. These trucks require close attention by the forklift driver to prevent degrade.

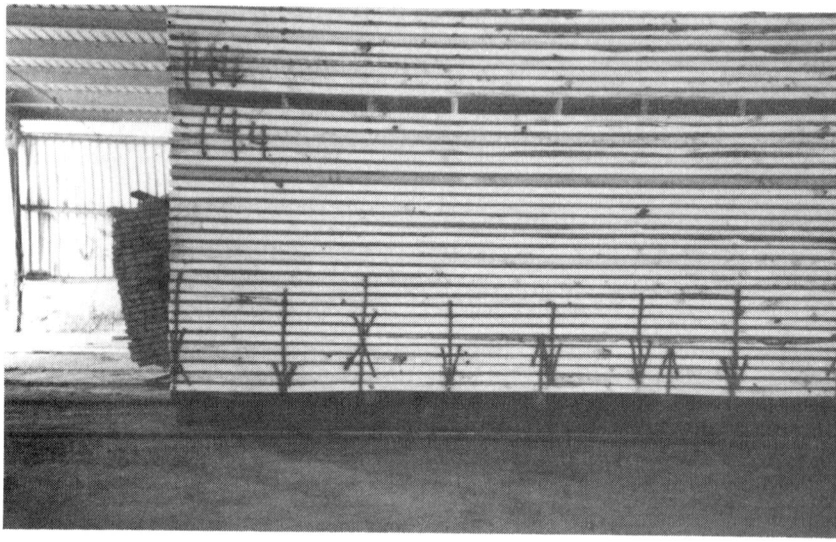

Figure 4.8. Mark a few units to visualize how loads are distributed to the trucks. Arrows pointing down are the forces imposed by the sticks, arrows pointing up are the truck supports. Note bottom layer distortion where forces are unopposed.

the kiln. This is best done at the stacker in transfer operations, at the kilns in others. Try Vekton™ bushings rather than roller bearings for your truck wheels. They roll more easily and require less maintenance. Although the manufacturers claim you do not need to grease these bearings, you will get longer life from them by doing so.

Whatever truck design you have, personnel must take the time to align bunk supports under the sticks. Illustrate this to your people by painting arrows pointing up at each truck support with arrows pointing down at each stick. If the points of these arrows do not meet, you are losing lumber to "kiln kink." This shows up in the dry sheds and at the planing mill. Do not make the mistake of thinking only a few boards are lost by poor truck placement. Each edge board represents an entire layer, and several layers are usually affected.

Keep units aligned top to bottom in the kiln so they do not lean over. Units that are offset from each other upset airflows by varying the width of the side plenum chambers.

Transfer operations stack their lumber directly onto kiln trucks at the stacker. They follow the same guidelines for use and placement of trucks. Since the entire stacked truck may be moved great distances, these mills often have trouble with trucks shifting out of position. This is particularly

Figure 4.9. Poor truck setups that alter the width of the plenum chambers and affect airflow uniformity top to bottom and along the kiln length.

troublesome on units that have a 2-foot gap between neighboring trucks. This always occurs in 10-foot, 14-foot, and 18-foot units. Mills that try to save on trucks see it in 16-foot and 20-foot units too. If the pull cable is attached to the truck or bunk support, trucks easily shift out of position. Avoid this by using the maximum number of trucks that will fit under the unit or by placing spacers between the truck gaps at the stacker. Some mills have the transfer operator cut and insert wooden spacers each cycle, while others fabricate steel spacers that travel with the trucks at all times. A simple solution that often works is to position the space on the end farthest from the cable attachment. This makes it harder for cars to shift under the load.

Crossouts

Forklift operations load kiln trucks at the kilns. In addition to positioning the kiln trucks, personnel must also insert and align crossouts (also called bolsters) separating the bundles. A few transfer operations unload with forklifts and also insert crossouts at the stacker.

Figure 4.10. Wooden spacer inserted between trucks to prevent truck shifting during transfer. This unit actually needs another truck to fill this gap.

Figure 4.11. Metal spacer to prevent truck shifting.

Figure 4.12. Place crossouts under every stick location. There is a truck support under every stick. This lumber is well supported to minimize drying degrade.

Crossouts should be good quality, straight, and unbroken. Since they serve to support the units above and rest on the units below, they must be uniform in size. Place one at each stick location. Failure to do this results in warped lumber in the top and bottom layers they contact. Arguments from forklift drivers that there is no room to maneuver forks with all these crossouts are unfounded. They simply do not want to handle the extra crossouts. This costs the mill money. When you make the drawing that locates stick guides at the stacker, also determine fork spacing. There is a spacing that works for all length combinations.

The most common material for crossouts is green hardwood 4 × 4s. Although inexpensive and readily available, they have a short service life when exposed to day-to-day kiln conditions. Pine 2 × 4s are often used, although they tend to fall down if stood on edge as the kiln is loaded. It is better to lay two on top of each other or nail two together. The main problem with both these crossouts is their tendency to break. Left in the system they degrade the lumber they contact.

Figure 4.13. Crossout breakage and poor lumber support from poor stacking practices. Scatter the boards on the top layer to help distribute the weight.

An alternative crossout material is square aluminum tube. Large and stiff enough to support lumber without bending, these crossouts are light and easy to handle. Although their initial cost is high, they pay back in extended service life and improved grade. One mill I know had been using green gum 4 × 4s as crossouts, with an average life of three months. Based on the purchase price for green gum and aluminum tube, we determined that if the aluminum would last one year it would pay back the investment. These crossouts have now been in continuous service for seven years with no loss of quality or usefulness. Unless accidentally damaged by equipment (they do not support forklifts driven over them) they have an indefinite service life. Heaven help us, though, if anyone ever figures out how to make hog pens from them.

Keep in mind that these crossouts never leave the kiln area. Kiln personnel place and remove them for each charge, keeping a rack handy for their storage. Pine 2 × 4s are used in the dry sheds. The 2 × 4 crossouts see only one trip through the sheds, since they are dressed along with regular production in the planing mill.

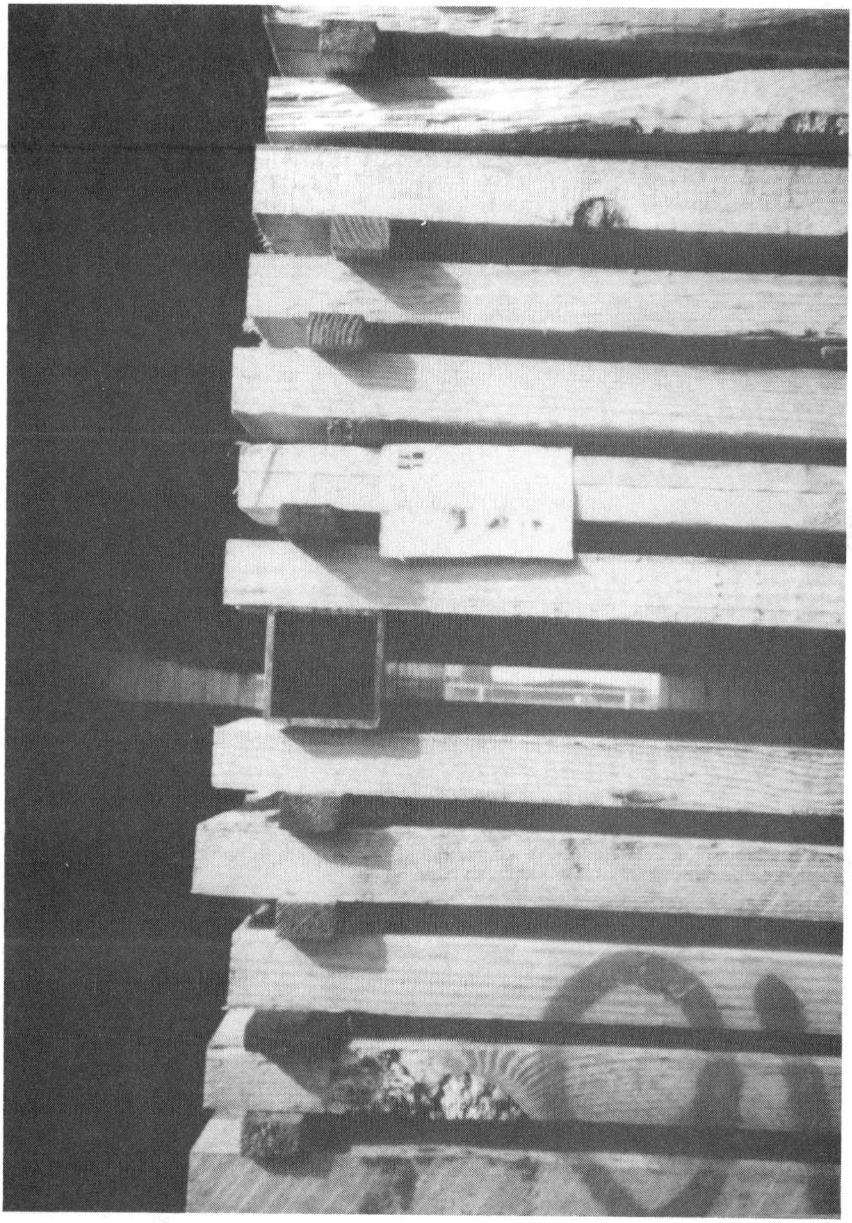

Figure 4.14. Tubular aluminum crossout. Photo courtesy of Union Camp Corporation.

Four inches by four inches is the maximum recommended size for crossouts. The smaller, the better, since a significant amount of air is lost through these spaces during the drying process. Some mills have found

Figure 4.15. Pine 4 × 4 and 2 × 4 crossouts. Note the large gap for air loss between the two units.

Figure 4.16. Damaged crossout causing lumber degrade.

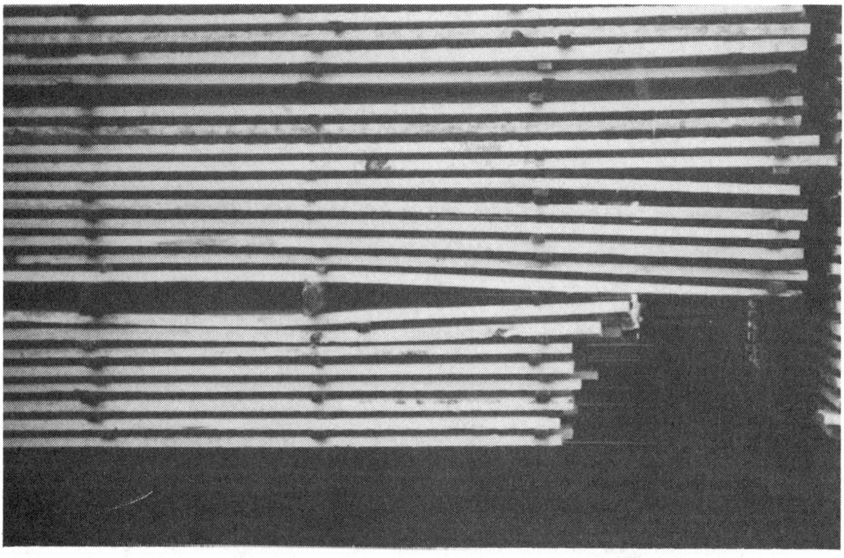

Figure 4.17. Insufficient crossouts in a rather unique lumber setup.

Figure 4.18. Lumber crossouts covered for an air velocity comparison check.

they can reduce crossouts to 1¾ inches by changing to wide, thin, high-strength steel forks. Due to the width of the fork you may have to eliminate two crossouts to have enough room, and this degrades a little lumber. A little diesel oil on the forks helps them slide in easier. Covering crossouts speeds drying by increasing the average kiln air velocity. It also narrows moisture distributions by eliminating the very dry boards surrounding the crossout space.

George Culp of Culp Lumber Company, New London, North Carolina, and I once ran a test on his new direct-fired kiln to determine the effect of closing these crossouts. We took air velocity readings with the crossouts open and found the average velocity was 975 fpm with a velocity range of 310 fpm. After pushing the lumber back out and covering the four crossouts with 1 × 4s, we reloaded the kiln and tested airflows again. Average velocities had increased to 1,118 fpm for a 15 percent increase in airflow. The range reduced to 250 fpm.

Blocking this space certainly helps, but I have not found a practical method for doing this in kilns with more than one crossout. George and I covered the crossouts with green 1 × 4s, but this was time consuming and not a realistic procedure on a day-to-day-basis. Other mills have taken

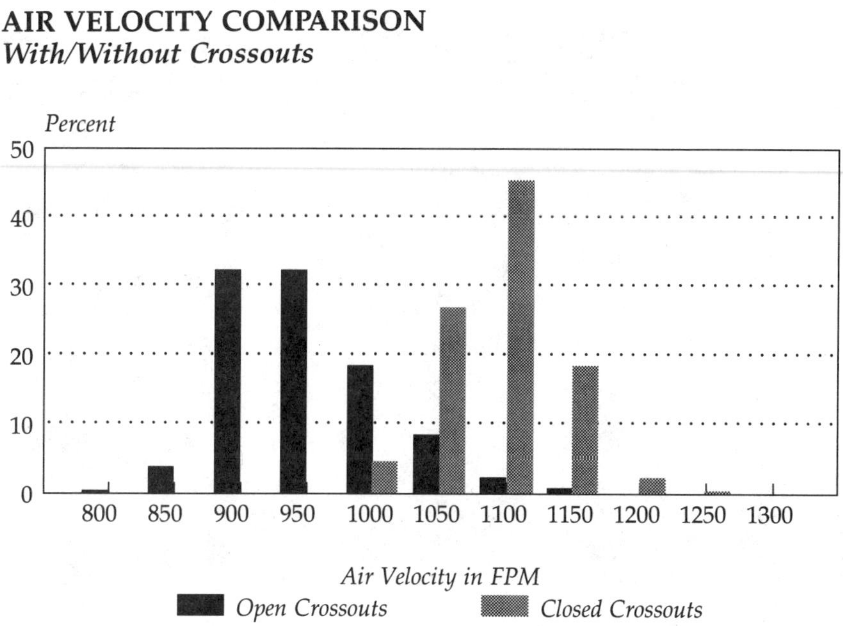

Figure 4.19. Shift in air velocity by covering the crossouts in Figure 4.18.

short lengths of 2 × 4s and placed them in the opening. Another mill stuffs short sections of firehose into the space. All work fine on crossouts you can reach from the ground. Once you must tote and climb a ladder to install blocks, however, the work and time involved outweighs the benefit in many operator's minds. Mills with only one crossout are lucky. This space is easily reached from the ground.

Choose among the above methods for the one that is easiest for you to implement. I like the firehose method, since you don't have nails and broken boards lying around. Since most mills usually have used firehose, it is also inexpensive.

I have seen another method of covering a single space that is simple and easy to install. At this mill aluminum strips $1/8$ inch × 4 inches × 10 feet hang horizontally on chains against the wall. Wherever the chain attaches to the strip, a clip made from heavy strapping is attached. This is the strapping normally used on chain rail cars. These clips slip into the stick spaces above and below the crossout space, holding the aluminum strips in place. The nice thing about this method is the lack of clutter on the floor, since the chains keep the strips against the wall and at a conve-

Figure 4.20. Crossout closing method that works well with one opening. The clips are formed from heavy strapping and insert into the stick openings above and below the crossout. The chain keeps them off the floor when removed.

nient height. They are quickly removed and inserted when changing the kiln.

Never leave crossouts on top of the top units within the kiln. Overhead baffles rest on them and leave a large avenue for air loss.

Important Loading Considerations

Always fill kilns to capacity. Plan ahead. Once you start loading it is usually too late. This is the difference between managing an operation and just running it. Your kilns were designed to hold a maximum amount of lumber; anything less is a lost drying opportunity. Shortloading a 68-foot double-track kiln costs you 1.5 percent capacity for each 2 feet of short track. If every charge is short 2 feet then you must dry an extra five charges a year to make up what you lose. If the charge is so short your end baffles do not contact the lumber, airflows become unbalanced and uneven drying results.

Figure 4.21. Shortloaded kiln. This gap represents a lost drying opportunity and has a negative impact on drying uniformity.

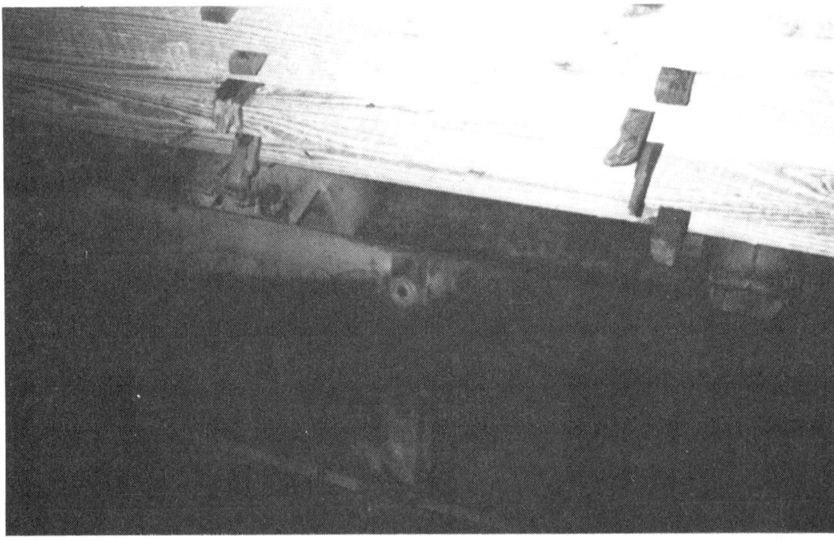

Figure 4.22. Set lumber units to the edge of the kiln truck so the floor baffles can contact the lumber.

When forklift loading, always set units to the very edge of the kiln trucks on the side next to the floor baffles. This ensures a tight fit with the baffles and eliminates the short-circuiting of air under the lumber.

Make certain that the total height of the loaded trucks are uniform from one truck to the next. Changes in height of one or two layers are not serious, but avoid drastic changes that result in gaps at the top where overhead baffles contact. This is a source of air loss. Baffles must contact the lumber to do their job.

Units of adjoining trucks must butt tightly end to end to avoid air-loss gaps. This is difficult at mills with poor length control, and results in the kiln operator trimming ends at the kiln to minimize the gap or just to close the kiln doors. Before trimming, however, determine that the lumber is indeed overlength and not just poorly even-ended at the stacker. This results in trimloss. If poor even-ending is the problem, these boards will usually push back into the units while loading. You will still have to trim boards to close the doors unless they are easily driven in by sledgehammer. However, if the exposed ends shatter, you might as well trim them. This is an opportunity to get the people in the sawmill or at the stacker involved in correcting the problems they cause. Once you identify the source of the problem, make sure those responsible see and understand the impact at the kilns.

Figure 4.23. Uneven load heights make it difficult for the top baffles to close the gap.

Figure 4.24. Poor loading practice. Note the crossouts left on the top of the unit, which leaves a gap for air loss.

Figure 4.25. Poor truck setups that leave large gaps between adjacent units.

Figure 4.26. When you set two units side by side on a truck, leave sufficient space between them for airflow to redirect as it passes through.

Kiln Loading and Pulling Methods

Just as there are two primary methods of kiln setups, there are two methods for loading depending on whether the system is transfer or forklift loaded. The method used at any particular mill is strongly influenced by the personal preference of the mill designer. There are advantages and disadvantages to both.

Transfer Systems

Transfer operations are labor and equipment intensive, requiring specialized equipment (transfer cars) and reasonably skilled personnel to do a good job. The primary advantage is the uniformity of stacking, which reduces the impact of crossouts and poor setups on airflow. This advantage disappears when kiln personnel fail to fill kilns completely and leave gaps within the kiln.

Use extreme caution when working around these operations, since cables used to pull lumber have a tendency to slip off or break. Another hazard is walking between units on the same track. I recall one instance in which a man was crushed between units as the kiln was being loaded. He was not discovered until after the load was dried because he was not a part of the regular workforce. Everyone who works with these systems quickly becomes attuned to the dangers. Give moving cables a wide berth.

Transfer cars have a tendency to become very temperamental in their operation, either through poor design or deteriorating maintenance. As an existing transfer system ages, the ends of tracks next to the transfer pits wear and are often patched. Concrete in the vicinity of these repairs chips out, offering little support. Truck wrecks become more frequent as slop wears into the system from the uneven forces exerted by moving large volumes of lumber by cable.

Here are a few guidelines for designing a reduced maintenance transfer system:

1. Rim all exposed edges of the transfer pits with heavy 3/8-inch-thick 3 × 3 angle iron. This stabilizes the edges and gives something for the cables to wear against.
2. Bury the rail in concrete up to the ball of the rail and slope track slightly in the direction of lumber movement through the kilns. Never build valleys between tracks for drainage. You will never be able to get a forklift in to correct truck wrecks. A flat, even surface ensures forklift maneuverability and is a lot easier to work around.

Figure 4.27. Transfer operator moving lumber onto the track for loading into the kiln. Note the mixing of different-aged lumber.

Figure 4.28. One of four hydraulic cylinders mounted on the corners of a transfer car. These cylinders stabilize the car as lumber is moved onto and off the tracks.

3. Use the heaviest cable that personnel can comfortably work with. A cable that is too light breaks easily and must be replaced often. Use the cable with the least stretch to reduce the tendency for lumber to overshoot the kiln. This creates additional work.
4. Install pneumatic or hydraulic cylinders at each corner of the transfer cars. Once the car is aligned with the tracks, these cylinders extend and stabilize the car in position, reducing the likelihood of wrecks.
5. Pull the dry lumber out of the kiln before pulling the green in. A few mills pull both at the same time and have higher than necessary transfer car and cable repair costs.

What is the best method to attach cables to the kiln truck? All methods are a compromise, but you have two basic choices. Either you attach it to the truck or you attach it to the lumber.

The most common method of attachment to the truck involves simply hooking the cable on the end of the truck support. This is not recommended since cables easily slip off the end and truck supports bend. This also imposes a twisting motion to the truck, which contributes to truck

Figure 4.29. Truck damage resulting from attaching pull cable to the ends of the bunk supports.

wrecks and damaged wheels and bearings. A better method is to lay a cable between the tracks with a hook on one end to hook the car and a loop on the other to attach to the winch cable. This requires two complete cable sets per track. Before green lumber is placed on the track a cable is laid between them. With the track loaded, this cable is used to pull the lumber into the kiln and is left in the kiln during the cycle. The second cable is laid between the green tracks and lumber loaded onto the tracks as before. When the dry lumber has been pulled from the kiln, that cable is moved to the green track and the process repeats. Individual units of dry lumber are easily moved onto the dry end transfer car with cables on the side. Any system that attaches cables to the trucks must make provisions to prevent trucks shifting underneath the lumber. This may be as simple as moving the space between trucks to the end opposite cable attachment. Here, you must place longer lengths on the end next to the door since 10-foot lumber almost always shifts their trucks. If you cannot control the sequence of lengths going onto the tracks then you must insert spacers between the trucks to prevent shifting.

Attaching cables directly to the lumber prevents shifting trucks, but usually results in mechanical damage to lumber and/or stickers. There are several techniques. Cables may loop around the corners about waist high and hook onto an end bunk support, pulling up on the truck. This breaks

Figure 4.30. Kiln pulling techniques that minimize lumber damage.

Figure 4.31. Kiln pulling techniques that damage lumber.

corner boards and sticks. Other mills drive bars or wedges into the end stick spaces. Then a cable is run between or beside the track and attached. In this case the ends of boards are damaged. Some mills intentionally notch the ends of lumber for cable attachment. The best system for this I have seen uses a heavy flat plate specially designed to insert in the sticker space. The bearing plate is wide to spread the force and reduce board damage. A cable is run under the truck and attaches to the plate. This works pretty well unless the device slips out. Fishing for a new cable under these trucks is not easy. It is doubtful that the people who designed most transfer systems ever ran one.

Forklift Systems

You will find two common methods for loading kilns by forklift. The first is set up similar to a transfer operation, in that one end of the kiln is

Figure 4.32. Pulling method for kilns that do not have a green or dry end. A piece of shafting is placed in brackets built on the ends of the trucks. Photo courtesy Union Camp Corporation.

designated the green end and the other end is the dry end. The dry end usually has a cooling shed. These systems must move empty trucks from the dry end back to the green end for loading. Often you will see a transfer system on the green end with a forklift removing dry lumber and returning trucks to the stacker. These systems either push in and out by forklift alone or use cables in conjunction with the forklift. Both systems are cumbersome and require precise maneuvering by the driver. They also put wear and tear on the trucks.

The loading system I prefer has no designated green or dry end of the kiln. A double-track kiln loads green lumber on the track at one end of the kiln and the other track at the other end. This permits access to both sides of either track for loading. To push in, a bar is placed into notches built into the end trucks, which face each other when the kiln doors are opened. The forklift operator then backs up to the first line of green lumber and pushes. As he pushes the green lumber in, the dry lumber pushes out. The bar is moved to the next line and the process repeats. This is very fast and efficient, allowing you to turn the dry kiln in 15 minutes or less. Truck damage is minimal, since trucks spend their entire lives

Figure 4.33. The forklift backs up to the green lumber and begins pushing. Photo courtesy Union Camp Corporation.

Figure 4.34. Once the dry lumber is out on that track, the shafting is moved to the other track and the process is repeated. This method reduces lumber and truck damage. Photo courtesy Union Camp Corporation.

on the track. There are two things to keep in mind, though. First, there can be no spaces between trucks within 20 feet of the end where the bar is positioned. Otherwise, trucks shift out of position as the load is

Figure 4.35. As long as the lumber is fully supported at the sticks, you can stack lumber in imaginative ways to accommodate different length combinations.

pushed. Second, it is often necessary for the lift to pick up a bundle of green lumber before pushing to increase traction. This is particularly important in the winter.

CHAPTER GUIDELINES FOR REDUCING VARIABILITY IN DRYING

1. Select lumber for each charge that is as even aged as possible. Dry your oldest lumber first.
2. Always fill kilns to capacity.
3. Use 4-foot kiln trucks with three supports spaced on 2-foot centers.
4. Place sufficient kiln trucks under each unit of lumber such that every stick has support. Make sure the supports align with the sticks.
5. If you use crossouts to set up your trucks place one at each truck location. Discard any damaged crossouts that affect uniform lumber support.
6. Cover crossout openings to reduce air loss through these openings.
7. Never leave crossouts on top of units going into the kilns.
8. Bring lumber to the outside edge of the truck on the plenum side to permit good floor baffle contact.

9. Keep the total height of loaded trucks uniform for good overhead baffle contact.
10. Butt the ends of units tightly together to reduce air loss between them. Align units top to bottom to improve airflow uniformity top to bottom and along the kiln length.
11. Adopt a kiln pulling method that reduces damage to lumber.

CHAPTER 5

Final Loading Considerations

> It is a funny thing about life; if you refuse to accept anything but the best, you very often get it.
>
> Somerset Maugham

Now that lumber is pushed into the kiln, we become more interested in certain design features of the kiln. Are we getting the full potential from our baffles? Do we lose too much heat through the doors or elsewhere? The answers to these questions have a significant impact on the quality and speed of our drying.

Everyone who runs a kiln sees, and should inspect, the inside of their kiln as it is loaded. However, most operators position their baffles and close the doors, never asking if their equipment actually performs its intended purpose. Have you ever entered your loaded kiln with the fans blowing and verified your impressions?

Kiln operators sometimes take a lot on faith, assuming that the kiln manufacturers have designed the right equipment to do the job. Never assume. The next time you start your kiln, take a moment to look around. Load the kiln as usual, positioning baffles and closing the doors. Lock out the heat system, turn the fans on, and close the vents. Now, walk down both plenum chambers and look. Take an air velocity meter if you have one. Do the baffles stay in position? Are they performing their intended purpose? How much air are you losing through gaps in the baffles or lumber? If you have crossouts, how much air is lost here? Do you feel cold air coming in around the doors? Do you see any light around the doors and vents? Is everything as tight as it could be? Make a note of the items you find amiss, and repeat your inspection with the fans blowing in the opposite direction. Correcting these problems goes a long way toward improving your drying operation.

The Baffle System

Baffles are hinged metal doors that are positioned against the lumber to close gaps between the lumber and floor, fan deck, and kiln doors. Their sole purpose is to direct all available air through the lumber. Baffles that do not contact the lumber or which move out of position during operation are not doing their intended job. Inspect your baffles each time you load the kiln. You have to handle them anyway, so pay attention to how well they are working.

Baffles must be durable and capable of taking abuse. Baffles constructed of 16-gauge steel with edges broken for stiffness are good. They tend to be a little heavier to handle, but stand up well in service. Avoid aluminum baffles unless you have severe corrosion problems. They simply do not stand up to wear. Aluminum overhead and floor baffles also require additional weight or restraint to keep them in place. Again, this is not apparent unless you inspect your kiln with the fans on.

Figure 5.1. Properly positioned overhead baffle with the end baffle folded with its edge against the lumber.

Figure 5.2. Properly positioned overhead baffle with the end baffle folded flat against the lumber side.

Overhead Baffles

Overhead baffles must fully contact the top of the lumber and extend to, but not over, the edge. These baffles smooth the transition from horizontal airflow out of the fans to the vertically aligned plenum chambers.

Baffles that extend over the edge restrict airflow through the top layers. Indeed, I have seen overhead baffles that literally envelope the top three to six layers, resulting in an engineered wet spot. This step is often taken to compensate for lumber units that are chronically overdry at the top. The real problem stems from poor control over heat and airflow distributions. Correct this at its source by multizoning to control heat variation top to bottom in the kiln. During your inspection you may be surprised to discover that these baffles will "sail" or float above the lumber. This reduces air velocity.

If the overhead baffles contact the lumber a few boards back from the edge, you will find the edge boards overdry and warped. The top edges of units also present a nonuniform profile to airflow, which creates turbulence. More serious, as the lumber dries it shrinks and the overhead baffles slip between the top board gaps. When these baffles are raised,

Figure 5.3. Lumber too short for baffle to contact load.

Figure 5.4. Varying load height reducing the efficiency of the overhead baffles.

Figure 5.5. Repair damaged overhead baffles immediately.

boards dislodge and fall. They are lost from the system due to breakage or excessive warp and counted as downfall. The safety hazard to the operator is real. I know: I was once knocked to my knees as a dislodged board glanced off my shoulder during an inspection. You learn very early to stay close to the outside walls as you raise baffles. A few mills construct elaborate mechanisms designed to protect the operator from falling boards. The real problem stems from improper baffle placement and the poor drying conditions that result. Correct the problem and enjoy the benefits in better drying and reduced degrade.

Remember, if the height of the lumber in your kiln varies over two layers then you are losing too much air over the top of your units. I have entered a few kilns to discover overhead baffles dangling well above the top layers of lumber. This is a sorting, stacking, or management problem.

Never leave crossouts on the top where they will hold the overhead baffles off the lumber. Some mills do this to stabilize the load and reduce dislodged lumber. Others do not recognize the harm in this practice, assuming the overdried top layer is lost anyway.

Repair damaged baffles as soon as possible after discovery. Large holes introduce variation to the drying system and affect drying uniformity. The most common source of damage results from failure to raise baffles when changing out a charge. Many mills compensate for this potential oversight by tying all overhead baffles to a single winch-and-pulley system, raising all baffles at once. I do not recommend these systems since gradual, uneven wear results in baffles with varying clearance. Undetected cable failures or an extra-tall unit will still rake off top layers and destroy baffles. The only difference is that you can blame it on that "@#$%&! cable" rather than poor memory or inattention. Pull each baffle individually to assure proper clearance. You do not need to wipe out too many baffles before you remember they are there, particularly if you must do the repairs.

IMPACT OF POOR BAFFLING
Missing Top Baffle

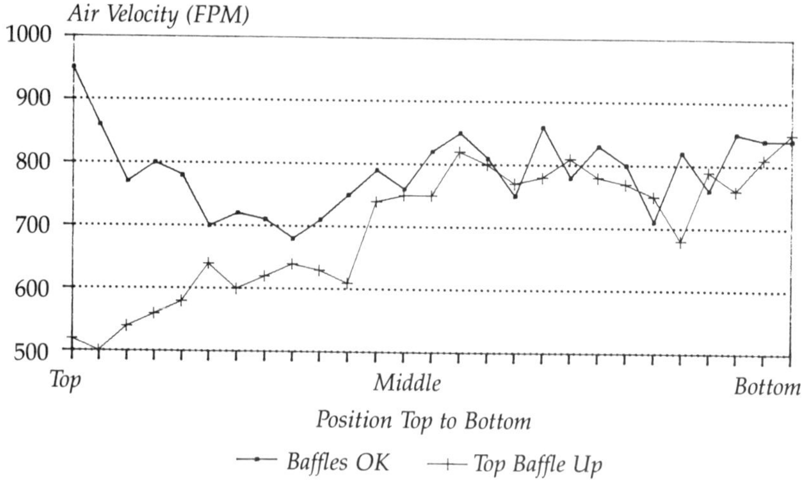

Figure 5.6. Variation in air velocity top to bottom in a kiln with and without the overhead baffle. Note the significant impact on the top one-third of the load.

Figure 5.7. Gap left by concrete floor baffle.

How much does keeping the top baffles up affect your drying? Figure 5.6 illustrates the change in air velocity off the second track with the baffles down compared to with the top baffle up.

Floor Baffles

Floor baffles close the gap between the floor and the bottom course of lumber. There are two main types in general use: poured-in-place con-

Figure 5.8. One method for closing the gap between the lumber and concrete floor baffles.

crete and hinged metal baffles. I prefer metal baffles, because they completely close this gap and do a better job of reducing air loss under the lumber.

Concrete baffles seem to be a carryover from early conventional kiln designs. These kilns were built with line-shaft fan systems and dried lumber with low temperature and air velocity. Drying times for 2-inch pine varies from 72 to 96 hours. Stationary overhead baffles and no end baffles were normal. If you still have one of these kilns it is possible to dramatically increase their airflow by installing hinged overhead and end baffles. At one mill running a conventional triple-track kiln we increased average air velocities from 175 fpm to 345 fpm by simply installing hinged overhead and end baffles. They also had concrete floor baffles.

The main problem with concrete baffles is that they cannot be moved. In order to provide clearance for the passage of trucks these baffles are installed with a 3-inch gap between them and the truck. Combined with sloppy lumber setups, I have seen this gap stretch to 9 inches or more. A large volume of air moves through this space and affects drying efficiency. To find out how much, measure the air velocity coming through the gap and compare it to that coming through the lumber or crossouts. Close the gap by using metal baffles that rest on the concrete and slide against the

Figure 5.9. Hinged metal floor baffle with retaining bar.

lumber. Attach handles to aid in placement and reinforce their edges with angle iron or bend (break) them for stiffness. A properly designed concrete floor baffle has its top surface even with the top of the truck and these metal baffles will contact the bottom board. The idea here is to close the gap without restricting airflow through the lumber. You may have to make modifications from the flat design to achieve this goal in your kiln.

Hinged metal floor baffles give the tightest fit. The main objection I hear is that, like overhead baffles, they are subject to damage and wear. Properly designed and cared for, however, they last a long time. The benefits of their use far outweigh their repair cost. The floor baffle design should have the baffle contacting the bottom board with enough slope on it to prevent its being flipped back by the airflow. Look for laid-down baffles when you pull your kiln and determine why that baffle fell. Common causes are boards or sticks overlapping the edge of the truck and warped baffles. When the baffle is held at a near vertical angle they fall easily

Figure 5.10. When you open your kiln after a charge, check to see if any floor baffles have fallen down. Investigate the reason for those that do.

when the air catches them. To correct this position lumber just to the edge of the truck. Replace warped baffles. Some mills elect to install restraints to keep them from flipping over under any circumstance. They require extra handling when pulling the kiln. If you have one of the new, super-high-velocity kilns, you may have no choice but to use them. Whatever works for you is the correct approach. Just make sure your floor baffles are doing their intended job.

I have seen floor baffles that varied from 2 x 12 lumber, to building siding, to corrugated water conduit simply rolled into place and held by a brick. None were designed into the kiln but were simply added as an afterthought. Get rid of them and install floor baffles specifically designed for your kiln. The payback is faster, higher-quality drying.

Figure 5.15 illustrates the impact of a missing floor baffle on air velocity through the lumber immediately above it.

Figure 5.11. Floor baffle held in place by end baffle. Edge board is sticking out from the side of the truck and holds baffle at a steep angle. Without the end baffle the floor baffle would fall.

Figure 5.12. Repair damaged floor baffles when discovered.

Figure 5.13. Metal floor baffle, which must be slid against the load. If one person must move this baffle in position it reduces the likelihood that it will be positioned correctly.

Figure 5.14. Floor baffle constructed of 2 × 12s.

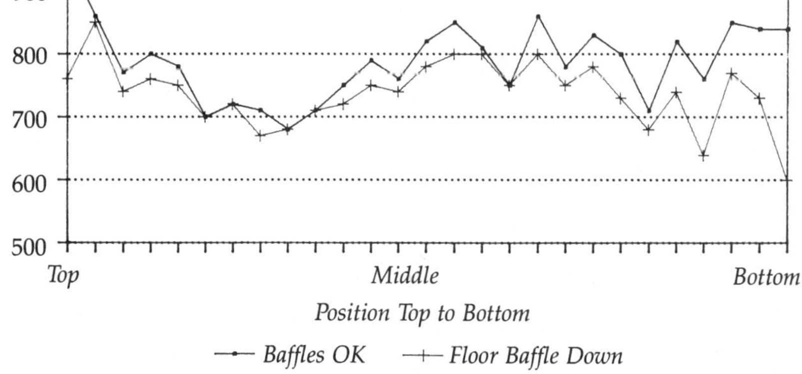

Figure 5.15. Variation in air velocity top to bottom with and without floor baffles. As expected, the impact is felt nearest the bottom.

Vertical End Baffles

End baffles prevent air from flowing around the ends of lumber next to the doors. Unless they securely close this gap they are useless. Most end baffles are bifold although single and trifold baffles are found.

Single-fold baffles are found at those mills that have made a firm commitment to always fill their kilns to capacity. They benefit from maximum kiln throughput and improved drying, because these baffles do not block airflow through the last two feet like bifold and trifold do. They are easy to hold in position, but they leave little room for error.

Bifold doors are the most common and accommodate variations in load length up to 4 feet. In fully loaded kilns most operators fold the extra 2 feet of baffle flat against the lumber. If they determine lumber dryness by metering the ends next to the doors, they must compensate for the higher moisture content of the end. If the baffle angle-iron hinge lies against the lumber, the shrinking unit will bind it. This dislodges overhead boards, which may strike the kiln operator as the baffle is opened. The alternative is to angle the outer baffle edge against the last stick position and secure it there. This avoids blocking the last 2 feet of lumber, but opens a large hole at the top for air loss.

Manufacturers maintain that trifold doors accommodate variations in load length up to 8 feet. I question this in high-velocity, high-temperature kilns. These baffles expose a significant surface area to the airflow and I have never seen one with sufficient restraint to keep it in place. Once the fans are turned on, they move about at will. In addition, they suffer from the same shortcomings as bifold doors.

Inspection of your baffles with the fans blowing often reveals glaring problems with positioning and restraint. When placed into position they must contact the outside edge of the lumber, ideally resting against the last stick. Never fold the baffle against the end. This looks like you are doing something, but once the fans are on they flatten against the inside of the door.

The key to the effectiveness of any of these baffles is adequate restraint. The baffles must remain in position, tight against the lumber, and block all air movement in their direction. The restraint must be firm, but flexible enough to adapt to changing loads. The spring restraints supplied by the kiln manufacturers are little more than glorified screen door springs. New ones have a nasty habit of snapping back in your face. They never seem long enough to reach the spot you need to hook them. Fortunately, or unfortunately, they lose their spring tension quickly and stretch out. Then . . . well, you get the idea.

Figure 5.16. A single-fold end baffle at a mill that has not quite made the commitment to fully load its kilns.

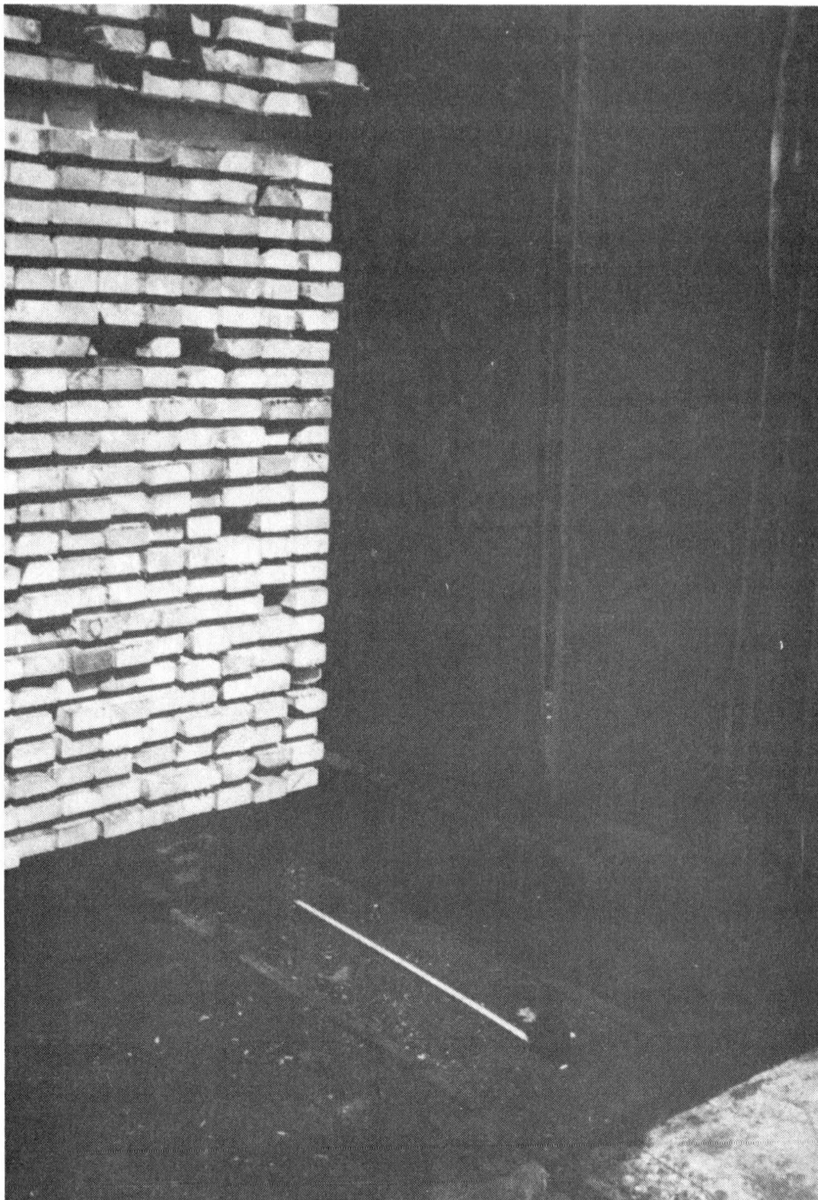

Figure 5.17. A properly positioned bifold end baffle.

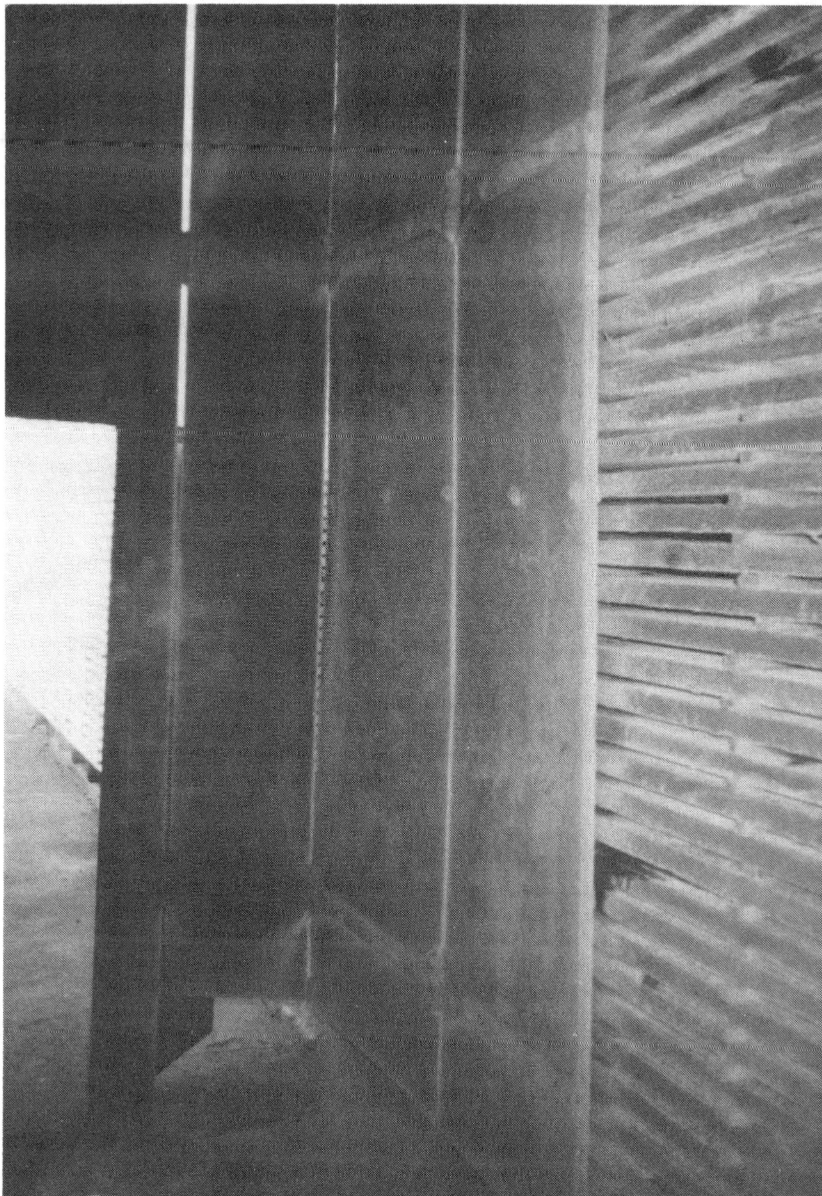

Figure 5.18. A trifold end baffle folded against the sides of a fully loaded kiln. Airflow is blocked in the end 4 feet of lumber.

Figure 5.19. End baffle improperly positioned against the end of the unit. With the air blowing toward the viewer, this baffle flattens against the inside of the door. Notice spring restraint.

Figure 5.20. Another end baffle positioned against the end of the unit. The chain restraint offers no resistance to airflow.

Figure 5.21. Proper end baffle placement with secure chain restraint.

There is a better way. This idea is the brainchild of Jim Brown at Union Camp Corporation, Seaboard, North Carolina. Jim uses a piece of 3/16-inch or 1/4-inch chain about 12 feet long for each baffle. He attaches it to the baffle halfway down its length to distribute the load. He hooks the other end onto a bracket inside the center door mullion, or wraps it around an extended board, stick, or crossout after the baffle is pulled tight into position. He uses a small chain binder to pull up the slack, but this is not necessary. The method is safe, simple, inexpensive, and—best of all—permanent.

Figure 5.23 illustrates the impact on airflow when an end baffle blows back. This data was taken 5 feet from the end at the third stick opening. The impact is greater at the first and second stick openings.

Kiln Doors

Kiln doors act as a movable, insulated, solid barrier to separate the internal kiln drying conditions from the outside. A remarkable number of mills seem to lose touch with that fact. Door panels are knocked out,

Figure 5.22. Method of chain attachment at the door mullion using a chain hook welded on the inside.

IMPACT OF POOR BAFFLING
Missing End Baffle

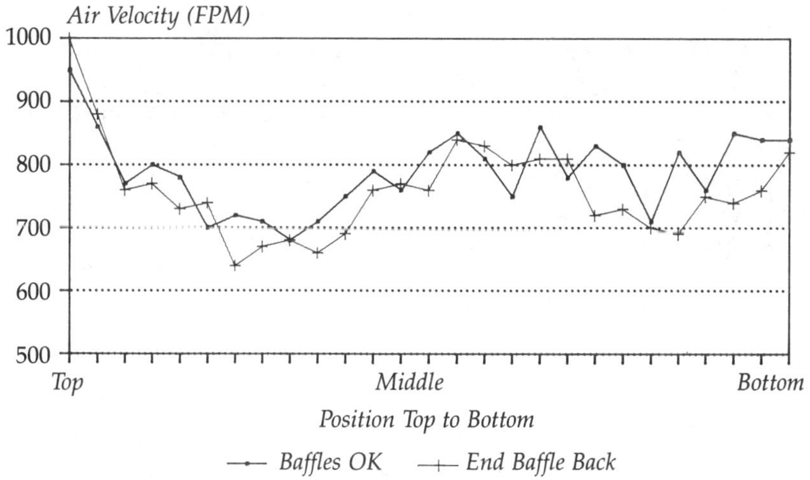

Figure 5.23. Variation in air velocity top to bottom in a kiln with and without an end baffle. Velocity taken in the third stick opening from the end. The effect increases toward the end stick opening.

Figure 5.24. Replace end baffles when damage is discovered.

frames bent, hold-ins broken. Repairs take months, if they are done at all. Keep in mind that the only reason we really bother to construct a kiln at all is to gain control over the conditions that regulate rapid, quality drying. If we lose that, then we defeat our purpose. Holes and loss of seals introduce cold, unconditioned air to the kiln environment—air that is neither sensed nor considered with conventional drying controls. Repair any damage to doors as soon as possible to maintain control.

Figure 5.25. Poor door closure.

Anyone who runs a kiln for even a short while soon wonders about all the steam that escapes around the seals of even a new door. Conventional tadpole gasketing has a very short service life due to the mechanical damage it sustains during normal handling. Alternative materials, like firehose, are more resistant to mechanical damage but break down quickly in the kiln. Older kilns are locked into some sort of gasketing material for life and should follow the manufacturer's recommendations.

Manufacturers must be challenged to develop sealing materials with built-in mechanical protection. Until then, fiberglass rope wrapped in fiberglass cloth seems to work best in high-temperature kilns.

New kilns, however, have another option. Well-designed kilns can avoid using gasketing altogether in favor of a metal-to-metal seal. This gives the best seal possible, but is highly dependent on the quality of design and fabrication to achieve matched surfaces. The first few charges seem to steam excessively around the doors as surfaces conform to each other. Over time, you can rely on a good seal without worry about damage and replacement. There will be leakage, but it will be far below conventional gasketing methods. Never attempt this on an older kiln; there is simply too much distortion to be successful.

To increase the effectiveness of a metal to metal seal, run a large bead of silicone sealant around the perimeter of the door facing where the door rests. Cover this bead with waxed paper and close the door. This will fill any surface irregularities.

With the doors closed and the kiln empty, inspect the inside of the doors. If you can see light, then you have identified a leak source. If it is at the top, check to see that the door is on the top hooks. These hooks slope down toward the kiln to use the door's weight to help hold it tight against the jamb. If it is on the hooks, but there is still light, check the bottom hold-ins. These occasionally fill with dirt and hold the door off the top hooks. They must allow free movement up and down. Finally, check the side mullions and their hold-ins. They must be snug against the side pins.

The spot where the tracks run under the doors has always been a problem. The common response is simply to notch the bottom of the door. This weakens the door and leaves an opening for heat loss. In addition, high-velocity, moisture-laden vapors stream out of this opening, eroding the exposed track. Complete deterioration often occurs in 6 to 8 years. The alternative is lift-out rail sections under the doors. Pioneered by Frank Cook on the West Coast, this innovation is occasionally seen in the South. Until a few years ago I had not seen a successful installation. The secret to success is in the foundation. It must be solid and perfectly aligned. Both the kiln and loading aprons must settle evenly or be rigidly aligned at the doors. Without notching, the integrity of the doors is maintained and tracks do not erode. An added benefit is that they act as truck stops when the kiln is running, preventing the forklift driver from accidentally pushing lumber into the doors while loading the next charge.

Erosion at the bottoms of doors affects not only tracks but the concrete sills. In new kiln installations and when replacing sills, always rim the

Figure 5.26. Lift-out rail section at bottom of door. Since the door bottom need not be cut to pass the rail, this improves the door seal and strength. In this position the rail section acts as a stop to prevent lumber from being pushed into the kiln door.

bottom sill area in angle iron. This results in a more uniform seal and, as such, reduces erosion to the point where it is not an issue. As a temporary measure, stuff gaping holes with burlap bags or cover them with belting cut to shape.

Door carriers should allow a single man to lift and move the kiln doors easily. This is a function of the design skill of the manufacturer. Being of slight stature, I quickly learned whose doors I could move without strain, and whose I could not. There are also a few I cannot move, period. Keep door carriers greased on a monthly basis (insist—no, demand—that new door carriers be supplied with grease fittings at all pivot and rolling points). In the full lift position, the bottom of the door must clear the track. If it does not, then something is wrong with the design. It is up to the manufacturer to correct it. I have seen kiln operators strain to pull the door out at the bottom as he tries to move the door to the side. Aside from the obvious toll on the body, the carrier wheels twist on their axles, causing premature carrier failure.

Doors are known to slip off their carriers and fall for any number of reasons. This is rather surprising to the operator and he is in danger of

Figure 5.27. Door sill deterioration covered with old belting.

Figure 5.28. Severe door sill deterioration, which must be rebuilt.

serious injury. OSHA requires that all doors be equipped with restraining brackets at the top to prevent such an accident. These brackets tend to become bent out of position over time, particularly if you have recently replaced your doors or done major repair work. Keep their purpose in mind.

It is because of poor design and operator frustration that recent years have seen the introduction of barn-style doors. These doors are split into

Figure 5.29. An example of a barn-style door.

two halves and hinged at the sides. Due to their size, their hinges must be well designed not only to carry the weight, but to swing easily without distorting the door or the jamb to which it is attached. Alignment is critical to satisfactory service. Their major drawback is that they must clear the tracks at the bottom as they swing out. This leaves a large gap, which must be closed in some manner. This usually involves some sort of gasket or belting arrangement with unknown long-term sealing problems. Any barn-door design must adequately address all the seal, alignment, clear-

Figure 5.30. Ground-level access doors are mandatory and handy when placed at each kiln corner.

ance, and maintenance issues without compromising the primary purpose of a kiln door: the separation of the outside from the internal kiln environments.

Access Doors

Ground-level operator access or "escape" doors are another OSHA requirement. I have never known of an instance where they were used in an emergency situation, yet it is comforting to know they are there. I can assure you that, faced with the prospect of being barbecued, I would fabricate my own door in very short order.

These doors are very handy for entering a kiln to check air flows, change wetbulb socks, or troubleshoot problems. Most operators open them to help cool the kiln before entering. Having an access door at each corner of the kiln is handy. Aside from the safety aspect, they allow you to pass quickly from one end to the other. These doors are best located on the end corners or side wall rather than on the larger doors. Those located on the doors force you to unlatch the baffles before entering. This becomes awkward when trying to check air velocity, because you must relatch the baffles from the inside to get an accurate reading.

A few manufacturers are placing another access door on the end wall above the main doors. This permits ready access to the fan deck for inspection, maintenance, and repairs. If you have ever had to squeeze through a fan deck trap door or climb between the overhead coils on the side plenums, you will appreciate this feature. The obvious advantage to the mill is that, having eliminated the ordeal of getting to the fan deck area, maintenance is kept at a higher level.

CHAPTER GUIDELINES FOR REDUCING VARIABILITY IN DRYING

1. Overhead baffles must contact the top of the lumber along the entire kiln length. They must not overhang nor extend back from the edge.
2. Floor baffles must prevent all air loss under the kiln trucks.
3. End baffles must remain in position when the fans are turned on and close the gap between the lumber end and the doors.
4. Replace damaged baffles and restraining devices immediately upon discovery.
5. Maintain kiln doors and their seals to reduce air infiltration into and out of the kiln.

CHAPTER 6

Dry Lumber Handling

Our chief want in life is somebody who will make us do what we can.
Ralph Waldo Emerson

Once lumber is dried and pushed from the kiln we must allow sufficient time for it to cool and equalize. Paragraph 161.4 of the SPIB 1977 rule book states that "Kiln dried lumber shall be adequately cooled after removal from kilns before dressing and final moisture content testing." How long does it take before lumber is "adequately cooled"? That depends on how badly the planing mill needs the lumber.

The longer you can let your lumber sit under shelter before dressing, the fewer problems you will have with it later. When lumber is first removed from the kiln, each piece has what is known as a moisture gradient. Have you ever watched the dial on a resistance moisture meter as you drove the needles slowly into a just dried board? At first the moisture reads low, but it gradually increases as the needles go deeper. This difference in moisture content between the surface and core is the moisture gradient. In high-temperature lumber this gradient can be very wide, with the surface being bone dry to a core approaching 20 percent or higher.

As the lumber sits in the cooling shed the moisture migrates from the high-moisture content (MC) core to the low-MC shell. Allowed to sit long enough, the moisture evens out. If you drive your needles into the board now, you'll find little difference between shell and core. The lumber has equalized.

But we have a planing mill foreman who just has to have the lumber that is in the kiln *now.* He cannot wait. He paces the floor moaning and groaning, worrying that the kiln will not be off before he runs out of dry lumber. He glances over toward the kiln, sees you opening the doors for your first look, and rushes over. "Is it dry? Is it dry?" he asks impatiently. After you complete your checks, you reluctantly conclude that he can have it. (You actually found a few units that weren't quite ready. Under less pressure, you might have run the kiln another half hour or so to be sure.) He is jubilant, rushing to the lift or transfer operator to impress upon him the importance of getting that lumber to the planer "RIGHT NOW!!!"

Figure 6.1. Improperly placed bunk supports for dry lumber. Degrade and trim loss is assured.

Figure 6.2. Proper positioning of bunk supports under the sticks; however, there are not enough of them to reduce lumber damage. This lumber needs four supports.

A week later the customer to whom this lumber was shipped calls and complains about excess moisture. The plant manager calls you and the planing mill foreman into his office to find out what happened. He turns to the planing foreman, who explains, "He"–meaning you–"said it was

Figure 6.3. Inadequate and improper bunk support placement. This lumber also needs four supports.

dry. We didn't bother to check it because you know how hard it is to check hot lumber. Besides, we would have had to shut down the planing mill because we were running out of lumber." The manager turns to you. "Well, I did check the lumber in the kiln and from my experience it seemed to be dry," you begin. "We had a couple of units that were a little high, but I've never had a problem with them before. I just don't understand what happened."

What did happen? Think about the purpose of a planer. It is designed to plane off the surface of the lumber to give you a uniform, smooth dimension. Since it removes the surface, it is removing the driest material from the board. There is nothing left to offset the high moisture content of the core. When the lumber does equalize, it will do so at a higher moisture content than it would have.

Let your lumber sit, under shelter, for a minimum of 24 hours. I prefer two to four days. Most transfer car systems are designed with extensive track and shed space behind the kiln for this. Forklift operations often

Figure 6.4. Inadequate and improper placement of bunk supports. There is a real danger of this lumber toppling over.

have a cooling shed over the dry end where lumber is allowed to set overnight. If not, then lumber must be removed to the dry sheds as soon as it is pushed out of the kiln. Hot lumber must not sit in the rain. The dry surface will soak up a lot of water very quickly. If you run out of shed space, place protective film on top of units to help them shed water.

In the dry sheds set up your supports as you would for stacked green

Figure 6.5. Nonstandard method of conditioning and equalizing lumber. Keep lumber under shelter to control moisture pickup.

lumber (chapter 2), placing sills so that the lumber does not span over 4 feet. The lumber must rest on the sills so that the sticks align with them. Crossouts must align with the sticks at the sill locations to eliminate distortion. Unlike green lumber, dry lumber is not flexible and breaks if not evenly supported. Keep dry lumber off the ground where it might pick up moisture, rocks, or grit.

Figure 6.6. Well-organized rough dry lumber storage. Photo courtesy Union Camp Corporation.

REFERENCES

1. Southern Pine Inspection Bureau, *Grading Rules–1977 Edition*. Pensacola, FL: SPIB, 1977.

CHAPTER 7

Fan Systems and Airflow

> Things may come to those who wait, but only the things left by those who hustle.
>
> Abraham Lincoln

The purpose of the fan system is to deliver heat to, and remove moisture from, the lumber surface. Uniform airflow is important for even drying, because areas of the kiln that have higher velocity dry at a faster rate than those where it is slower. Fast airflow through the coils of a steam-heated kiln transfers more heat to the lumber. The variation in heat delivery due to airflow was the driving force toward the development of multizone dry kilns, but we shall discuss that in more detail later.

Fan systems for high-temperature kilns are divided into two distinct categories. Direct-drive fans mount the fan directly onto the motor shaft with the motor and fan assembly mounted within the kiln environment. Shaft-driven fans mount the fan on a separate shaft. There are two variations on this. One method extends the shaft out the side wall, where the motor is mounted and belted to the exposed shaft end. The second uses a shorter shaft, which is belted to a motor mounted on the roof and is known as a ridge-mounted fan drive.

Direct-drive Fan Systems

The earliest high-temperature drying systems were built using direct-drive motor systems. Kiln manufacturers found that by mounting individual motors and fans within the kiln they could increase air delivery over what was possible with conventional line-shaft systems.*

---

* Note: Line-shaft fan systems have a single-fan shaft extending the length of the kiln and mounted in the overhead fan deck area. One end of this shaft protrudes through the end wall into a room that houses the fan motor, steam valves, and controls. Fans, mounted at intervals along this shaft move air in one direction through an arrangement of baffles. Air direction is reversed by simply reversing rotation of the motor. These systems do not generate sufficient airflow for efficient high-temperature drying.

Figure 7.1. Direct-drive fan system fitted with adjustable fan prop.

Motors mounted within the kiln are exposed to high heat and humidity conditions, which shorten the life of ordinary motors. At one mill running two kilns with seven fans each, we logged 18 to 21 motor failures per year. We were able to justify conversion to externally mounted motors based on rewinding, repair, and downtime costs, even though we purchased new motors for the system. It is dismaying to watch multiple motor failures occur on startup, knowing the work involved to change them.

The primary source of failure is not so much the heat but the high humidity of the kiln. Most failures occur immediately on startup. The reason is simple. Between charges or after a long idle period, the motor windings become relatively cool. As the kiln is loaded, moisture begins to condense on the motor housing and windings. This is a particular problem in kilns that share common walls. Often the insulation and vapor barrier qualities of the common wall are gone. Moisture vapor passes freely between the kilns. When the power is turned on, the motor shorts to ground. Prior to a startup, we once tested the continuity of each leg of the motors in the kiln. All were shorted to ground. In a panic, we turned on the heat to evaporate the moisture we found dripping from the motors. We lost no motors on startup.

A few manufacturers take advantage of this technique by leaving a 24-volt charge on the windings while the motors are idle. This keeps them warm and prevents condensation. This precaution is highly recommended for direct-drive fans.

Changing motors in these kilns challenges the stamina of the best millwright. Unless your kiln has access doors mounted on the gable ends, motors must be hoisted through a manhole. They must be manhandled and dragged along the fan deck (often corrugated), over and under steam pipes, sprinklers, and structural supports. Then they must be lifted into place. Just thinking about it makes me tired. If you have several kilns sharing common walls, this becomes very interesting—particularly if an adjacent kiln is running. Steam billows into the fan deck area, condensing on everything. Visibility—forget comfort—suddenly becomes important. To do anything, you must either shut down the offending kiln or at least set its fans to blow away from your kiln. Again, it is clear that most of the engineers who design these systems, and the people who purchase them, never, ever, ran one.

Shaft-driven Fans

Shaft-driven fans seem like the answer to every kiln operator's prayers. In general, they are. But you must be very careful when considering the different designs available. Some create additional problems with respect to uniform airflow.

The most uniform system, in terms of airflow, centers the fans directly along the centerline of the kiln. Why this makes a difference has to do with the fan manufacturers' recommendations on how far away from a fan you should begin turning the airstream. They recommend three fan diameters' distance. Most kilns must compromise on this a bit; but as long as the fans are centered, fan performance is similar regardless of fan direction. Performance differences increase as fans move away from the centerline. Some kiln manufacturers, in the interest of economy, move their fans way off-center to reduce shaft and bearing costs. Resist this temptation to save a few bucks. Airflow measurements in these kilns show significant variations in airflow patterns between the two fan directions. This affects final moisture distributions. Insist on fans mounted centerline of the kiln.

External motors are mounted either at the eave or on the ridge of the kiln. Each is belted to the fan shaft. Ridge-mounted, also called "through-the-roof" fan systems, are a recent development; the jury is still out on long-term performance.

Figure 7.2. Cross-shaft fan system fitted with adjustable fan prop.

This is how "through-the-roof" fans work: The fan is mounted on a short shaft (36 inches or so), centerline or slightly off-center in the kiln. There are only two bearings. The sheave end of this shaft is boxed in and

Figure 7.3. External motor for the fan system in Figure 7.2.

enclosed from the shaft to the kiln roof, where the motors are mounted. Drive belts pass through this protective enclosure and connect the motor to the fan shaft. They are less costly because of the short shaft (36 inches

Figure 7.4. Ridge-mounted cross-shaft fan system. Note the belt shrouding, which blocks a portion of the fan arc.

versus ±18 feet) and reduced bearings (2 versus 4 or 5). However, there are several potential problems.

First, the belt enclosure is very large (18 inches to 20 inches wide) and

Figure 7.5. Fan motors and drives mounted on the kiln ridge and belted to the internal fans.

positioned close to the fans. This blocks airflow through a portion of the fans' arc. As each blade passes this obstruction it unloads, causing it to flex and vibrate slightly. At normal rpm each blade flexes 865 to 925 times per minute leading to metal fatigue. We do not know the long-term impact this situation has on blade life. We do know that it adversely affects fan efficiency by acting as an obstruction to airflow.

Second, it is very difficult to tell if a fan has thrown its belt. Ammeters on the motors help, and allow you to compare current draw between all the fans in the system. They also allow you to detect slippage due to gradual belt deterioration. Monitoring and recording amp readings on a regular basis must be part of your preventive maintenance program. Most mills rely on visual inspection to discover fan losses. The location of the motors and the shrouding around the belts makes this difficult to accomplish on a regular basis. It is recommended that if belts are worn on one fan, you should replace them all. This avoids the constant uncertainty one-at-a-time replacement brings and results in better system performance. This repair is better scheduled rather than imposed by sudden failure. Ridge-mounted fan systems require a high level of operator concern to maintain them properly.

Finally, once a belt failure is detected, changing the belt becomes tricky. You must shut your kiln down and allow it to cool or wait for the beginning of the next charge to make repairs. This costs you more downtime. These shrouds are not vapor tight, so expect more frequent belt failures due to the action of heat, moisture, and wood chemicals. These belts must be tensioned more to overcome thermal expansion, and this hastens belt failure.

The most common external motor fan system mounts the motor at the eave of the kiln. This design has seen the longest service. A long shaft with multiple bearings transfers power to the fan, ideally mounted central to the kiln. These systems are more expensive due to the extra shaft and required bearings. Properly designed, they have the best service and performance record. Belt replacement requires only a short interruption of service.

A good design mounts the shaft on a large tubular steel base extending from the fan to the outside edge of the kiln. The tube must be heavy enough to maintain bearing alignment as well as transfer its load to the supports without distortion. Tube supports must control not only the vibration about the shaft axis but also the fan thrust. Heavier shafts increase stability and reduce the number of required bearings. Split bearings are preferred, eliminating the need to strip the shaft to replace them. Early designs placed a special bearing, known as a thrust bearing, next to the fan. Its purpose is to absorb the thrust of the fan and maintain its position. This bearing should be next to the sheaves to control belt alignment.

Motor Horsepower

When I first became interested in high-temperature drying, in the late 1970s, the recognized minimum air velocity was 700 fpm. It did not seem to matter that even with this as the minimum, actual velocities might range up to 1,100 fpm. My own and other studies have demonstrated the significant impact these variations have on final moisture content. Our purpose is to maximize airflows to their greatest extent while minimizing the variation.

There has been a great deal of controversy over the horsepower required for quality high-temperature drying. The size of your motors is strongly influenced by such factors as lumber thickness, sticker thickness, the presence of crossouts, plenum width, and door height. The right horsepower for your application is different from someone else's. Unfortunately, there are no hard and fast rules to determine what that horsepower should be. That is based on the prior experience of the kiln manufacturer.

Not long ago 15 hp and 20 hp fans were the recommended sizes for

Figure 7.6. Kiln eave access door for easy inspection and maintenance.

high temperature kilns. Today 25 hp and 30 hp tends to be the norm. Higher horsepower and their additional air volume tends to narrow the variation in air velocity top to bottom within the kiln. So do wider plenum chambers. I lean toward the higher end since high horsepower combined with variable speed motor controls allow you to tailor your kiln to your needs and save energy, too.

Recent research at Mississippi State indicates that drying times can be reduced by high-velocity/ultra-high-temperature drying.[1] These studies examined velocities to 2,000 fpm and temperatures to 300° F. Achieving these velocities in a high-production kiln has proven difficult using existing fan technology. The fan laws predict that in order to double the present air velocity, you must increase horsepower eightfold. Successful application of this process will doubtless require a kiln very different in appearance from today's models.

Variable-speed Motor Drives

We have known for decades that reducing airflow after achieving fiber saturation point (FSP) is one way to reduce energy usage without affecting drying speed. The ability to reduce fan speed by varying line frequency has been available for many years. Simple two-speed motors have been available for over 50 years.

Historically, the application of variable-speed motors required the interaction of an operator. Someone must be on hand to turn the switch that reduces fan speed. Furthermore, speed reduction can only be done after the lumber has passed FSP or significant extensions in drying time occur. To gain full benefit in energy savings we must reduce speeds on a consistent charge-to-charge basis. Change speed too soon and it costs money in extended drying time. Too late and you lose the full benefit of the technology. A pretty tall order for an operator who juggles several jobs at once and has no clear idea what is happening in the kiln.

I have visited several mills that had kilns built with two-speed fan technology. In no instance were the operators using the two-speed ability. When asked, the invariable reply is, "Yes, we experimented with fan speed at the beginning but stopped using it soon after the kilns were built. We couldn't see any benefit and it was a hassle to use. It's much easier to run one speed." The key words here are "hassle to use" and "easier." Something that is difficult to use, understand, or apply will not be used, regardless of benefit. I suspect that after one or two instances of reducing speed too soon any operator would give up in frustration.

The computer has had a dramatic impact in this area – so much so that I do not recommend installing variable-speed fans unless they are under automatic computer control. Not only will the computer slow fan speed automatically as programmed, it can respond automatically to moisture content or time as desired. Setting fan speeds becomes as simple as programming in setpoints for drybulb and wetbulb. Because of the computer, variable-speed motors will work. Together they give the lowest energy consumption and maximum return on investment.

Studies have shown that reducing airflow at FSP reduces fan energy costs by 40 to 50 percent. What is so magic about FSP? For most species the fiber saturation point is around 30 percent moisture content (MC). Above 30 percent MC water fills the cell cavities and its removal is by capillary movement of moisture to the surface. There it evaporates. This is a very rapid process and requires rapid movement of air to create surface turbulence and remove moisture-laden air. Below 30 percent MC moisture is held in the cell walls and its removal is quite different. It moves by the slow diffusion of water from high moisture content areas (inner portions of the board) to lower moisture content areas (the board surface). Lower air velocities will remove this surface moisture without affecting drying speed. Computer systems incorporating some method of detecting moisture content (TDAL, electrical resistance, load cells, and so on), can detect the FSP threshold and reduce fan speed automatically.

And the benefits do not stop here. It is now possible to maximize airflow by controlling your motors to a constant amperage. On startup cold, heavy kiln air is harder to move and requires more energy. As the air heats and becomes lighter, amperages fall off by 10 percent or more. Controlling to a set amperage allows us to reduce the initial strain of startup and increase airflows as air becomes lighter. This allows us to increase the potential of existing motors in those kilns or those lumber items requiring extra-high velocity. In some instances it allows us to use a smaller motor and still deliver enough air for drying.

We can use the motor as a brake to quickly slow fans for faster fan reversals. Although a minor benefit, this is a more efficient use of our drying time. On restart, its speed slowly increases to reduce or eliminate initial startup power surges. This saves belts and lessens the strain on windings, bearings, and sheaves.

Variable-speed fan technology is an old idea whose time has come. Poorly applied in the past, modern computers have opened up a whole new world of possibilities for their use. It works because the technology finally exists to make it work.

Table 7.1 shows a sample calculation for estimating energy savings from reducing fan speed.

Fans

On older kilns you may find fixed-pitch fans incapable of adjustment to increase air delivery. This locks the mill into fan outputs that may be inadequate for good high-temperature drying. These mills typically have low-horsepower motors, on the order of 7½ hp to 15 hp. These mills can do little to improve kiln performance or speed drying without completely rebuilding their fan systems.

Adjustable-pitch fans are available and recommended. They give you the option to increase airflow within the maximum potential of your existing motors. Keep in mind that for motors already operating at maximum amperage, changing to adjustable bladed props has no impact unless the fan efficiency of the new prop is greater. In general, more blades per prop tends to be more efficient. In most situations improvement in airflow requires a motor change along with the change in fan prop.

Fan System Maintenance

Before doing any work on your fan system, lock out the breakers so they cannot be started. Do not forget the main breaker. Safety is foremost in any situation, and this is no exception.

Table 7.1 ***Estimating Energy Savings from Reducing Fan Speed***

Assumptions:

1. 1 kiln with (7) 10 hp motors operating 8,500 hr/yr
2. Kiln variable-speed motor profile during cycle approximated as follows:
   - 16.6% of drying cycle at 100% speed (900 rpm)
   - 41.7% of drying cycle at 75% speed (675 rpm)
   - 41.7% of drying cycle at 50% speed (450 rpm)
3. The following measured data is representative of a typical kiln drying cycle:

	1st Period (16.6%)	*2nd Period (41.7%)*	*3rd Period (41.7%)*
Constant Speed	100%	100%	100%
amps @ 460 VAC	125 A	119 A	98 A
Variable Speed	100%	75%	50%
amps @ 460 VAC	125 A	60 A	38 A

Equation:

$$KVAH = (3 \times V \times I \times HR) / 1000$$

Where: KVAH = kilovolt-amp-hours
V = line to neutral voltage
I = line amperes
HR = hours operated

Calculation:

Total KVAH used at CONSTANT SPEED:

$$\begin{aligned} KVAH = \{&[3 \times (460/1.732) \times 125 \times (0.166 \times 8{,}500)] + \\ &[3 \times (460/1.732) \times 119 \times (0.417 \times 8{,}500)] + \\ &[3 \times (460/1.732) \times 98 \times (0.417 \times 8{,}500)]\} / 1{,}000 \end{aligned}$$

KVAH = 753,346 KVAH

Total KVAH used with VARIABLE SPEED:

$$\begin{aligned} KVAH = \{&[3 \times (460/1.732) \times 125 \times (0.166 \times 8{,}500)] + \\ &[3 \times (460/1.732) \times 60 \times (0.417 \times 8{,}500)] + \\ &[3 \times (460/1.732) \times 38 \times (0.417 \times 8{,}500)]\} / 1{,}000 \end{aligned}$$

KVAH = 417,283 KVAH

Total Savings Using Variable Speed vs Constant Speed:

SAVINGS = 336,063 KVAH or 45%

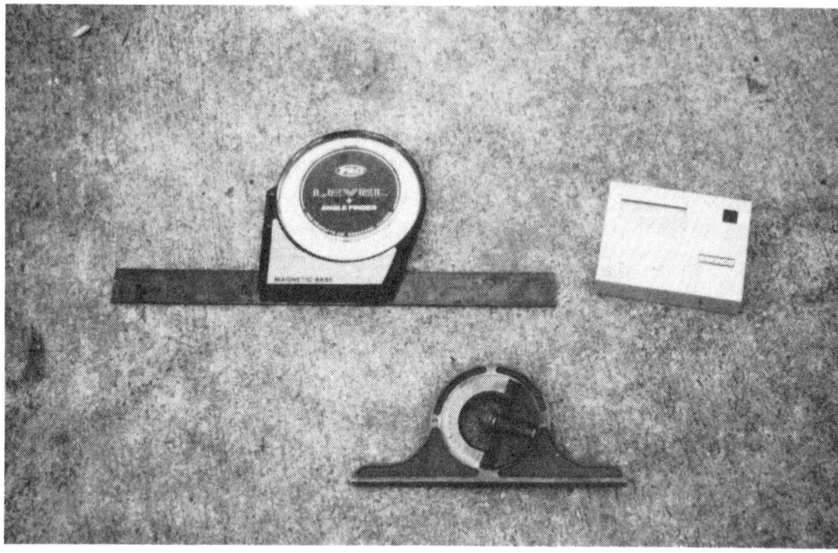

Figure 7.7. Collection of fan-pitch measuring devices, clockwise from left. Inclinometer, electronic angle finder, and machinist's level. The inclinometer is inexpensive and easy to use.

The worst part about doing any fan maintenance is getting to the fans. In older kilns it is necessary to hoist yourself through a trap door in the fan deck or squeeze between the overhead coils. Progressive kiln manufacturers recognize this problem and install gable end access doors with catwalk access in steam kilns, internal ladders and catwalks in direct-fired kilns. This encourages frequent inspections, as does the presence of fan wall bulkhead doors. Even when I know the fans are locked out, I am always reluctant to crawl through a fan to get to the other side. Manufacturers who place bulkhead doors between every other fan should be elevated to sainthood.

Grease bearings on all systems at least once a month under a preventive maintenance program. Direct-fired kilns require more frequent greasing due to the presence of abrasive ash in the air.

A few mills install copper or stainless steel grease lines that extend outside the kiln. This allows the kiln operator or maintenance man to grease the kiln while it is running. Others have automatic grease systems. I do not recommend any of these. Grease lines have a tendency to work loose, depositing grease everywhere but in the bearing. Grease also hardens in these lines and makes it difficult to tell what you are actually doing. Always grease at the bearing to verify that you are doing what you think you are doing. This forces you to enter the kiln and allows you to inspect the bearing for wear and schedule replacement if necessary. Inspect fans at this time for cracks, loose nuts, and excessive movement.

Use a quality high-temperature grease recommended by the kiln manufacturer or lubricant supplier. Never mix different greases in the same bearing unless approved by the manufacturer. Different greases are often not compatible and may react chemically with disastrous results. If fans do not have large retaining washers and bolts to hold them on the shaft, install them. Fans have a nasty habit of walking off the end of the shaft and through the gable end. They might even take a few neighboring fans with them.

Install ampmeters on each fan motor to monitor operation and assist balancing. As mentioned, recording fan amperages on a regular schedule helps detect slipping belts. It also alerts you to the increasing friction produced by gradual bearing failure. Fan outage lights are not dependable indicators of fan outage since they are tied into the starters, not the motors themselves.

Inspect fan blades for cracks. Examination of broken fan blades shows

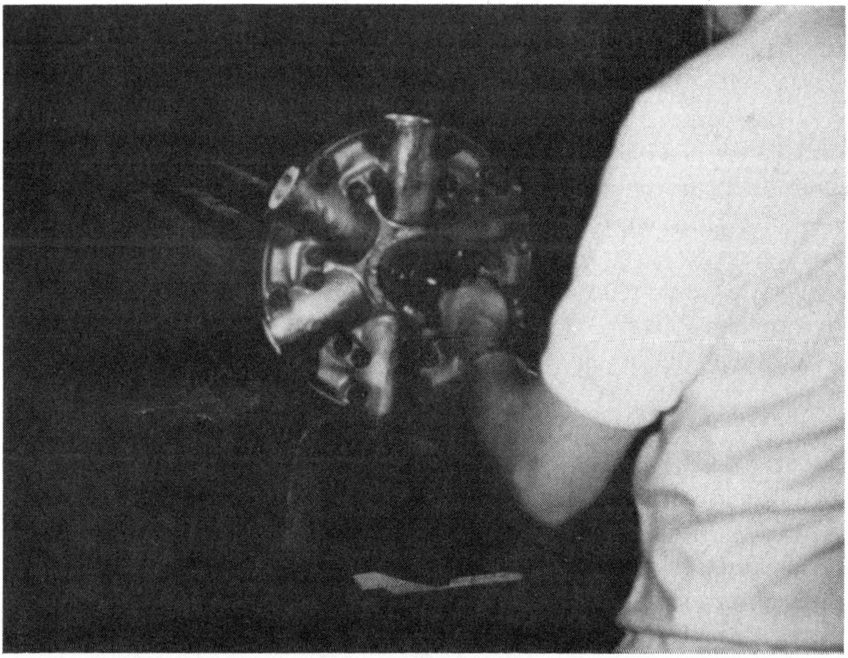

Figure 7.8. Removing a fan hub that has thrown its blades.

a tendency for them to break one-third of the radius from the hub. Indeed, it is sometimes possible to detect hairline cracks at this position upon close examination. Do this on a routine basis.

In new kilns check the torque on fan U-bolts, bushings, and retaining washers after each charge for the first week. After that check the torque twice a year. Be careful not to over torque. Nothing is more dismaying than taking one last nudge on the wrench and hearing a loud pop as the hub cracks. Never exceed manufacturer's specifications, no matter how little you feel you are tightening the bolts. Watch out for bolts with bad threads. They can fool a torque wrench. I like to use a wrench that allows you to dial up the desired force in inch/pounds and makes a snapping noise when the correct force is applied (inch/pounds = foot/pounds × 12 inches/foot). Avoid wrenches that force you to read a dial or have pointer indicators. They are inaccurate and slow you down.

Check clearance between blade tips and fan shrouds. Clearances should not exceed 1/2 inch nor allow the blade tips to touch. A broken shroud often slips down into the path of the fan blade, not breaking it, but slowly wearing it down. I have seen blades reduced in length by 2 inches or more without breaking the fans. Once repaired, replace the fan. Tip clearance is critical to efficient fan performance.

Keep a spare fan on hand in case you lose one. I would not bother to assemble it until you need it. Keep it in its box on a shelf and assemble it on the fan deck. I once assembled a fan on the ground because it was easier, only to discover that there was no way to get it upstairs.

Always keep spare belts on hand if you need them. Which belts should you use? Some manufacturers use matched sets of single belts. These belts give good service and are relatively inexpensive, although they are sometimes difficult to tension. I prefer multiple belts attached to a common back. However, excessive sheave wear will split them in short order, and they can also be difficult to install. Their advantage is that you do not have to worry about tensioning several belts that vary slightly in length. Another option is power bands, which are cogged belts running on a cogged sheave, not unlike an automobile timing belt. Properly installed, these belts do not slip. This is their primary advantage. But they are also expensive and require more effort to install. Sheave alignment is critical, and they also tend to be a bit noisier than regular belts.

While making your fan inspection look at other areas of the kiln. In steam kilns look for signs of leaks or unusual coil movement. Direct-fired kilns always seem to lose bolts from the distribution duct. Vent linkages slip and require readjustment. Look for damaged baffles along with signs of wall panel and structural steel deterioration. Keeping your kiln in good repair is important to maintaining drying quality.

Fan Operation and Reversing Times

When fans first start at the beginning of a new charge, there is a real danger of the kiln walls "imploding" or being sucked in. It is embarrassing to explain to your boss that a hole has suddenly appeared in what had been a solid aluminum wall. How does this occur? The answer involves the expansion of kiln air as it heats up after the doors are closed and its sudden contraction as it passes through the cold lumber when the fans start. All manufacturers make provisions for this in their control systems, invariably opening the vents just prior to starting the fans. This allows pressures to equalize. Some are manual and require the operator to remember to do this; *all* should be automatic. A good computerized system opens the vents a few seconds before starting the fans and only starts them if it receives a signal that they are indeed open. Once the fans are started, the vents immediately close. This reduces energy loss during fan reversals.

Older kiln systems brought all fans on together, creating a massive power surge. Later systems sequence fans individually or in pairs to reduce surges. Variable-speed drives bring all the fans gradually up to speed to eliminate surging and have the added advantage of reducing strain on the motor's windings, belts, and sheaves. This is the wave of the future.

Most mills reverse their fans every 3 hours. I occasionally encounter 2- and 4-hour reversal times. Some mills reverse only once, two-thirds or so through the schedule. Reversal periods appear to have little impact on drying and tend to be a matter of personal preference. Certainly, fewer reversal periods reduce your drying time by eliminating idle periods, but this is more of a factor with the old dial reversing timers. They require 3 or 4 minutes to complete a reversing cycle. Modern electronic timers and computer systems can complete a reversing cycle in less than a minute. The shortness of the cycle depends on how long the fans take to coast to a stop. Variable-speed fan drives shorten this further by using the drive as a brake, making a smooth transition from one direction to the other. An old friend of mine, an electrical engineer, always wanted me to "plug reverse" my fan system. In essence, this makes an immediate direction change without any coast-down allowance. Every time we met, he reminded me of his suggestion. All I could see in my mind's eye were squealing, smoking belts, twisted shafts, and blown motor windings. When he comes around, just smile and nod your head.

Getting the Most from Your Fan System

Obtaining peak performance from your fan system means both maximizing fan output and balancing fan output along the kiln length. Not only do we realize the full potential of available motor horsepower, but drying variation is reduced.

Direct-drive fan systems are not concerned with power losses from the motor to the fan. Shaft-driven systems have the potential to lose a great deal of power to improper belt tension. Belt maintenance is very important to fan efficiency. If you hear a high-pitched squeal or squeaking sound after the fans have come to speed, you are likely losing fan efficiency.

Before balancing a shaft-drive fan system, check belt tension and adjust it if necessary. Replace worn belts and sheaves to maintain drying efficiency. One way to reduce belt and sheave wear and smooth startups is to install airbags under the fan motors. This maintains flexible belt tension and reduces the belt squealing typical of fan startups. With better control over belt tension, the variation in fan motor amperages narrows. With our belts tight, we can begin balancing.

In many cases the angle graduations on fan blades and hubs are not accurate. This is a particular problem in darkly lit older kilns. Set the pitch of each blade using an inclinometer. This assures even pitch on the blades. Verify the pitch settings yourself, even on new installations. I prefer to use an inclinometer with a magnetic base and attach the 12-inch metal rule from a tri-square. The tri-square rule functions as a wide base and, for 6-foot fans, provides a handy gauge to mark one-third radius on the blade. Most fan manufacturers recommend checking blade pitch at a point one-third of the radius in from the blade tip.

Procedure for Setting Fan Pitch

To set fan pitch you will need a lumber marking crayon, an inclinometer with an attached tri-square rule, and a recording pad and pencil. If you are adjusting the blades you will also need a torque wrench with the right sized sockets. Take along sockets and an open end wrench to tighten bushing bolts while you are at it.

In order to get the maximum airflow your motors will deliver, you need information on fan amperage. With the kiln loaded for a charge, turn the fans on and record full load amperage for each motor while the kiln is cool. This is your worst-case load. Most operations maximize amperages with 2-inch lumber in the kiln. Keep in mind that motors work harder to push air through thicker lumber (i.e., fewer sticker openings). You may want to maximize amperages on your thickest item to avoid motor overload. To determine the rated amperage for your motors, look for the information stamped on the motor nameplate.

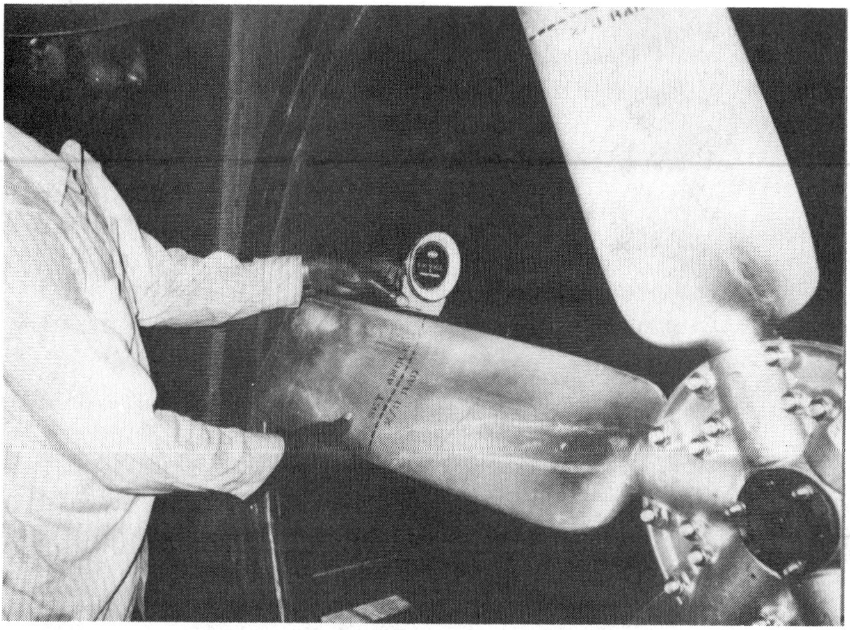

Figure 7.9. To check fan pitch, first adjust level of the fan blade so it is horizontal.

To document the impact of your efforts, measure air velocities before and after balancing. Check and record airflow through at least three stick openings in each unit of lumber plus the crossouts. Spot check air loss around the units.

You should refer to the manufacturer's instructions for setting fan pitch. In general, the following procedure works well:

1. Face the fan assembly with the drive shaft going away from you. Position the fan hub on your right.

2. Number the blades with your crayon and bring the first blade down horizontally in front of you.

3. Set the inclinometer on the top edge of the blade and level it to 0 degrees. Make a mark on the fan shroud level with the top edge of the blade. This is your horizontal mark.

4. One-third of the radius in from the tip of the fan you may find a dotted line. If not, use the rule to mark a line with your crayon at this distance. This is where you read the angle for setting pitch.

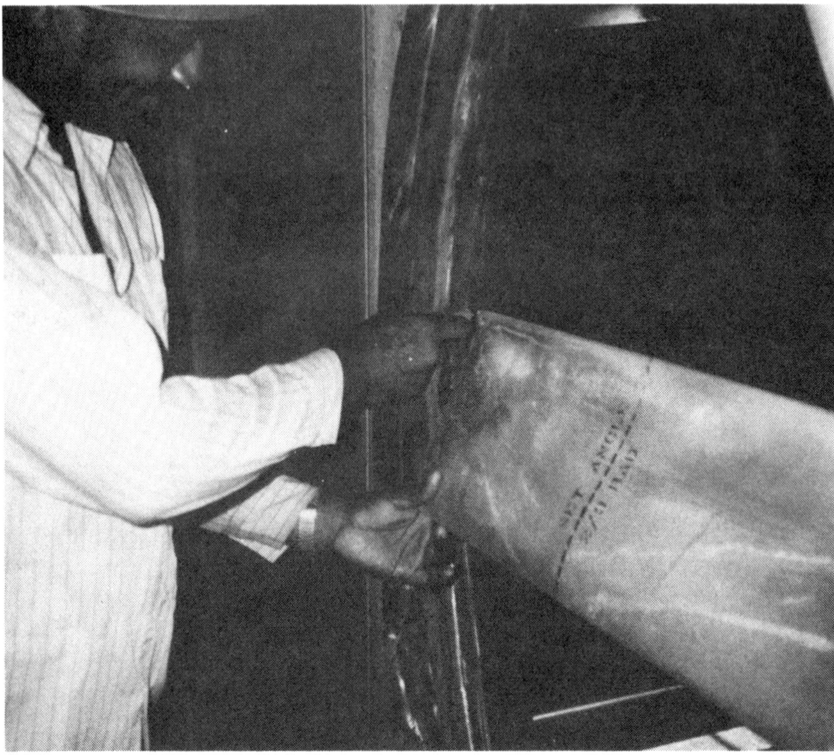

Figure 7.10. Mark the horizontal line on the shroud. Notice the manufacturer's mark for checking fan pitch.

Lay the inclinometer on this mark with the straightedge tight against the blade, while holding the fan blade horizontal with the mark on the shroud.

5. Read the angle or set the blade to the desired angle if you are adjusting the pitch. I always mark the angle on the blade and record all readings at the end of the survey to save time. Adapt the record sheet at the end of the chapter to record data for your operation.
6. Repeat steps 4 and 5 for each blade, being careful to number each fan in sequence.

Table 7.2 shows an example from an accelerated kiln with 7.5 hp fans and 4-bladed adjustable pitch fans.

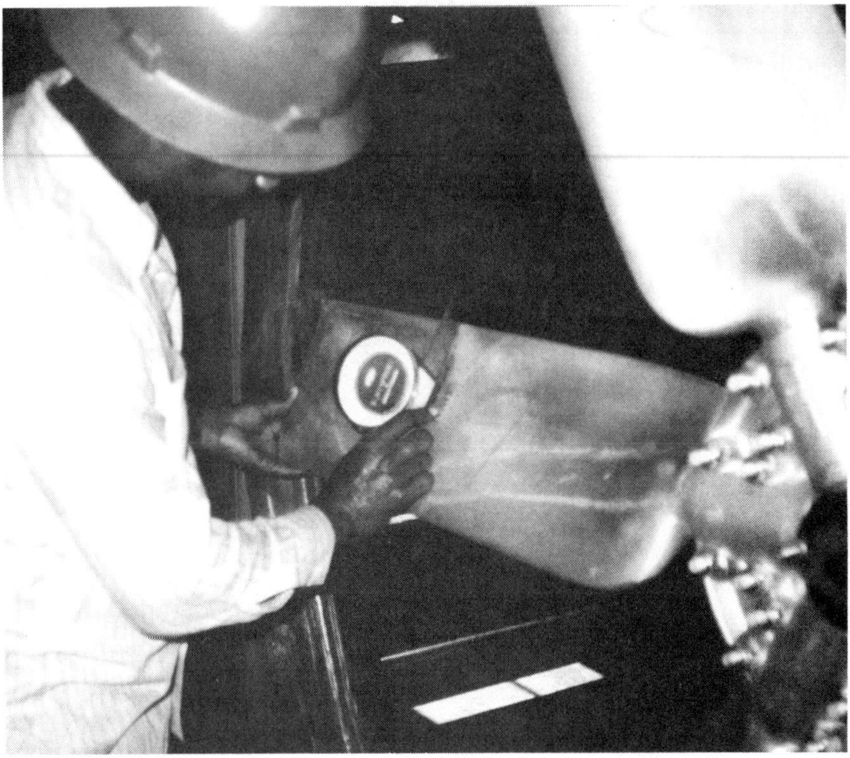

Figure 7.11. Lay angle finder on the blade at two-thirds of the fan radius.

These angles are measured from the horizontal axis. The manufacturers measure fan blade pitch setting from the vertical axis and would express this pitch as 14.3 degrees. Since the meter I recommend for checking fan pitch reads from the horizontal axis, I present the data in this form. Just keep the difference in mind when comparing your readings to manufacturers' recommendations.

Higher horsepower allows more fan pitch. For example, a 30-hp motor with a 6-bladed fan might accept 61.5 degrees from the horizontal, 28.5 degrees from the vertical. It pulls 34 amps.

Determining the maximum pitch your fans will accept requires some experimentation. I like to set several fans to different pitches, turn them on, and compare amp draw. In the example above we set fans to 74, 72, and 70 degrees and ran them. This gave us the information we needed to make a decision.

Table 7.2 ***Fan Pitch Settings and Motor Amperages***

Kiln number: 1 | Date: 3/6/86
Rated Moter Amperage: 15.5 amps 7.5 hp, 880 rpm | Average air velocity: 544 fpm | Range: 340 fpm
Notes: 7/4 charge, 2 units high on truck, one crossout, AS FOUND

	PITCH SETTINGS BLADE NUMBER							
Fan #	*1*	*2*	*3*	*4*	*5*	*6*	*Avg.*	*Amps*
1	76.0	76.0	76.0	76.5			76.1	10.5
2	76.5	78.0	76.5	77.0			77.0	10.8
3	77.0	77.0	77.0	77.0			77.0	13.0
4	78.0	79.0	77.5	80.0			78.6	11.8
5	73.5	72.0	73.5	75.0			73.5	12.5
6	77.0	77.0	77.5	78.0			77.4	11.8
7	76.0	77.5	77.0	76.0			76.6	14.3
8	78.0	77.0	76.0	75.5			76.6	14.3
9	67.0	68.0	66.5	67.0			67.1	15.3
10	77.5	76.5	77.0	77.5			77.1	10.5
					KILN AVERAGE		75.7	12.1
					MAXIMUM		78.6	19.3
					MINIMUM		67.1	10.5
					RANGE		11.5	4.8

Now let's look at Table 7.3.

So, what have we accomplished in Table 7.3? We increased the pitch by almost 4 degrees, which increased average amperage by 2.4 amps. We narrowed the variability along the kiln length, since fans turn the same rpm regardless of fan pitch. That is not all. Look at the average air velocity moving through the lumber. It increased over 30 percent from 544 fpm to 710 fpm with negligible impact on variability. This change reduced drying time by 14 percent—certainly a worthwhile effort.

Some people like to overamp fans on startup since total amperage drops off quickly as the air warms. I have heard recommendations to 15 percent above motor rating and it is certainly worth a try to increase output. Actual drops vary from 10 percent to 16 percent and are highly dependent on your situation. Compare the amperage drops with your present fan pitch by comparing cold startup amps to a reading taken one

Table 7.3 ***Fan Pitch Settings and Motor Amperages***

Kiln number: 1 Date: 3/6/86
Rated Moter Amperage: 15.5 amps Average air velocity: 710 fpm
Range: 350 fpm

Notes: AFTER FAN BALANCING, SAME LUMBER LOAD

	PITCH SETTINGS							
	BLADE NUMBER							
Fan #	*1*	*2*	*3*	*4*	*5*	*6*	*Avg.*	*Amps*
1	72.0	72.0	72.0	72.0			72.0	13.0
2	72.0	72.0	72.0	72.0			72.0	14.0
3	72.0	72.0	72.0	72.0			72.0	15.5
4	72.0	72.0	72.0	72.0			72.0	15.0
5	72.0	72.0	72.0	72.0			72.0	15.0
6	72.0	72.0	72.0	72.0			72.0	16.0
7*	73.0	73.0	72.5	73.0			___	___
8	72.0	72.0	72.0	72.0			72.0	15.0
9	72.0	72.0	72.0	72.0			72.0	13.3
10	72.0	72.0	72.0	72.0			72.0	14.0
					KILN AVERAGE		72.0	14.5
					MAXIMUM		72.0	16.0
					MINIMUM		72.0	13.0
					RANGE		0.0	3.0

*Fan number 7 would not accept the higher pitch and later failed. It is omitted from the comparison figures.

hour later. If you try this, add a little cushion for safety. Keep in mind that you must watch your system closely after you make this change to verify that fans do not overheat and drop out. Losing a fan has a tremendous negative impact on drying conditions. Do nothing that will affect system reliability.

Airflow Patterns for a Steam-Heated Kiln

The airflow patterns created within a steam kiln are influenced solely by the stacking and loading methods along with the efficiency and balance of the fan system. Unlike direct-fired kilns, airflow patterns are not influenced by the heat distribution system. We will discuss direct-fired

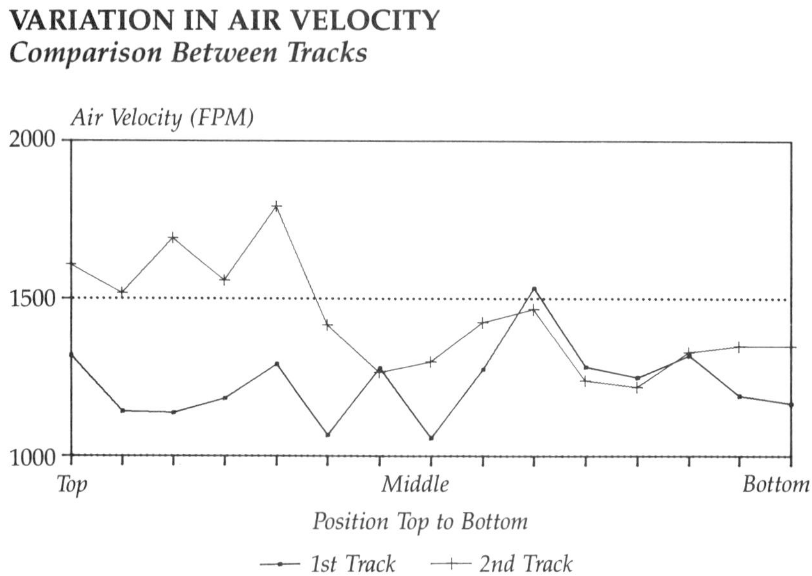

Figure 7.12. Comparison of air velocity top to bottom in a steam-heated kiln. Data shown from both tracks in a two-track kiln.

kilns and the influence of their heating system in the chapter on these kilns.

By taking firm control of our stacking and loading practices we can radically reduce the variability of airflow within our kilns. By balancing and maximizing airflow we reduce this variability further. Still, there are variations inherent in the kiln design that are unaffected by these improvements. There still remains the natural variation top to bottom in the kiln, and differences between tracks in two-track kilns. Figure 7.12 illustrates these variations in a typical kiln.

As you can see, top to bottom variation is greater in the air velocity coming off the second track. Top to bottom variation is smaller off the first track. It is apparent that there is a general migration of air toward the upper corner in the second track.

An interesting approach to balancing this airflow is the installation of plenum baffles in the outside walkway to split and control the airflow, a concept initiated by Gayle Bridges. Plenum baffles work on some kilns and not in others. I once installed these baffles in a mill that had two kilns, each radically different in design. The #1 kiln had a center coil that could

Figure 7.13. Plenum baffles installed in a multizoned kiln. Control sensors are mounted on the inside of the baffle for the top center coil and just below the baffles for the bottom center coil.

STANDARD STEAM KILN
Air Velocity Down Kiln Length

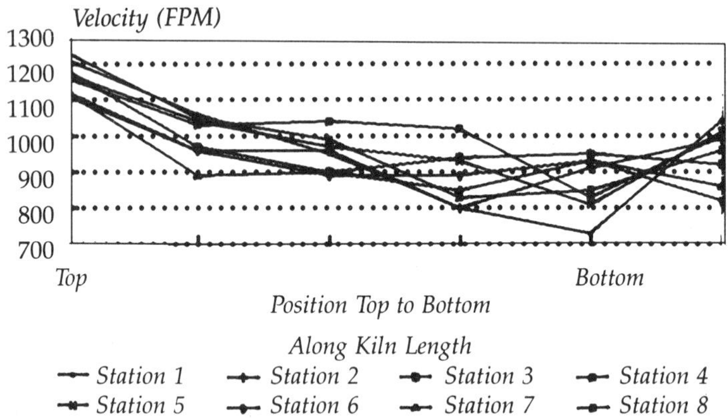

Without Plenum Baffles

MODIFIED STEAM KILN
Air Velocity Down Kiln Length

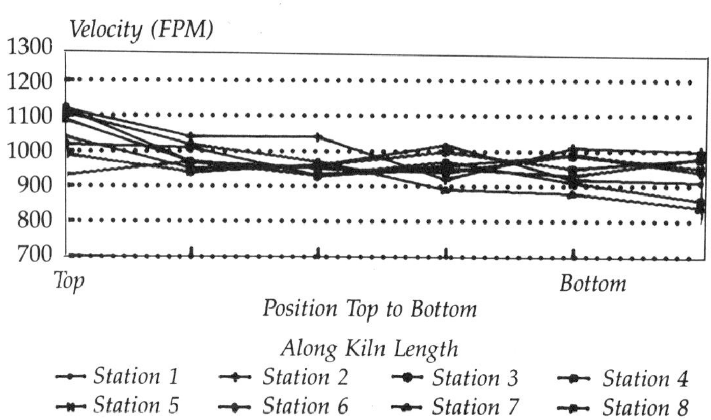

With Plenum Baffles

Figure 7.14. Comparison of air velocity down the length of the kiln in a steam kiln before and after installation of plenum baffles. Data taken off the second track.

not be split top to bottom into zones. Indeed, the center coils were oriented vertically. The second kiln was multizoned into two zones top to bottom on the center coil. After we installed plenum baffles in the first kiln we saw a significant drop in drying time and an improvement in grade. In my enthusiasm, I recommended these baffles to other mills and found that some mills got good results while others could see no impact. Meanwhile, the first mill installed the baffles in the second kiln and saw no improvement for that kiln, although they had seen and continued to enjoy improvements in the first kiln.

It appears from this experience that if you are able to control temperature top to bottom through zoning on the center coil, you will gain no benefit from installing plenum baffles. If you do not have control over the top to bottom heat variation off the center coil, you will gain approximately one-half the potential benefit of zoning the coil. If you later zone the center coil the plenum baffles lose their utility, although they cause no harm by keeping them installed.

Figure 7.14 illustrates how the air velocity changes top to bottom in a kiln by installing properly engineered plenum baffles. Although there was no change in average air velocity, the range was cut in half from 500 fpm to 280 fpm. I do not have reliable information from installations of plenum baffles in direct-fired kilns.

REFERENCES

1. Taylor, F. W. and P. H. Mitchell. "Can Drying Times Be Shortened Further?" *Forest Industries* (November 1987): 24.
2. Culpepper, L. G. "Coming Trends in Kiln Drying—What's New on the Horizon?" Reno, NV: West Coast Dry Kiln Association Meeting, May 1988. Published in 1989 proceedings.
3. Culpepper, L. G. "Application of Computerized Controls to Improve the Drying Operation." Seattle, WA: IUFRO Meeting, July 1989.

CHAPTER 8

Humidity Control

Success is a journey, not a destination.

Ben Sweetland

The humidity control system of a drykiln is probably the least understood part of the entire kiln system. We easily grasp how the air moves through the lumber and how heat enters the kiln. Yet, without humidity control, drying either slows to a stop or proceeds at too fast a pace. Both conditions have the potential to degrade lumber.

At a given temperature the air is capable of holding a maximum amount of water. Relative humidity measures the percent of moisture in the air up to a maximum of 100 percent. If the relative humidity is high, near 100 percent, the air does not readily accept additional moisture and evaporation slows. You notice this on hot, sultry summer days, when your skin stays moist because your perspiration cannot evaporate as fast as it is produced. At lower humidities the air readily accepts additional moisture and you remain dry and cool as moisture evaporates freely.

In a drykiln we manipulate temperature, airspeed, and humidity to control lumber drying rates. At any temperature and airspeed we can speed or slow drying by adjusting the relative humidity within the kiln—raising it to slow drying, reducing it to speed drying. There are some limits, however, to how fast or slow this can be done and still maintain lumber quality. In actual practice we do not measure relative humidity directly but use a method of drybulb (a simple thermometer is one) and wetbulb (a thermometer with a wet cloth over it). As water evaporates from the wetbulb, it cools. Slow evaporation cools it a little, fast evaporation cools it a lot. How much it cools in relation to the drybulb temperature is an indicator of relative humidity. The difference between the drybulb and wetbulb temperatures is known as the wetbulb depression.

Given enough time, wood within a drykiln controlled at a set temperature and humidity will stabilize at a certain moisture content. This moisture content is known as the equilibrium moisture content (EMC). When wood is at EMC the water in the wood has equalized with the moisture in the surrounding air; it neither gives up nor accepts any net moisture change. Tables are available that tell us the EMC of wood at different drybulb temperatures and wetbulb depressions. We control EMC conditions, and thus the drying rate, by maintaining drybulb and wetbulb conditions to a safe level within the kiln.

The humidity control system of the drykiln consists of drybulbs, wetbulbs, vents, and steam sprays. The vents exhaust excess moisture to the outside and draw in cool, dry air to maintain the desired humidity. A few high-temperature kilns are able to raise humidity conditions by spraying steam into the kiln environment. We will discuss these systems and how they work together to control drying.

Humidity Sensors

High-temperature drykilns use wetbulbs to control humidity conditions within the kiln. This method is simple, requiring only a temperature sensor (known as a drybulb) with a cloth sleeve or "sock" pulled over it. This sensor mounts above a water box in which the water level is controlled. The sock hangs into the water and stays wet by wicking moisture up from it.

Recently, other methods for determining humidity and EMC have been introduced. However, they are limited by present technology only to lower-temperature drying processes. EMC wafer systems allow us to determine EMC directly by measuring the electrical resistance of a known fiber sample. This system tends to respond slowly to changes in kiln conditions. Electronic relative humidity meters read humidity directly, but the extremes of heat and humidity found in these kilns impose severe service conditions for this type of equipment. Any decision to use alternative humidity-sensing methods must be on a trial basis until proven. For sheer simplicity and low-cost reliable operation, the wetbulb is hard to beat. I will continue our discussions using the wetbulb as our humidity sensor.

Number and Location of Wetbulbs

Until the 1980s it was rare to find a high-temperature kiln with more than one wetbulb. After all, what would you do with it? Most mills running high-temperature kilns never used wetbulbs to control anything. They simply monitored drying progress and perhaps used it as an aid to pull the kiln.

The first computerized high-temperature kiln initially had two wetbulbs, one on each side. We were certain that significant differences existed from side to side in the kiln and we would be the first to discover

Figure 8.1. Typical wetbulb installation.

and control it. To our surprise, however, temperatures rarely varied more than one or two degrees and we soon abandoned the extra wetbulb. For conventionally controlled steam-heated drykilns, one wetbulb is enough.

As computerized controls became more sophisticated, a few manufacturers began installing two bulbs again. Not to measure and control anything different, but as a backup in case one bulb failed. For example, wetbulbs occasionally run out of water for one reason or another. When they do, they become drybulbs and begin reading drybulb temperatures. Conventional controls do not sense that anything unusual has occurred and immediately open the vents to bring the temperatures down to setpoint. The kiln runs out of control until someone notices the problem and corrects it. In the meantime lumber degrades. A well-designed computerized system senses the problem and automatically switches to the other bulb for control. Further, an alarm prevents the operator from simply ignoring or overlooking the wetbulb loss. This improves equipment dependability and reduces one source of accidental lumber degrade.

Mount your wetbulbs midway of the kiln on the side walls, 6 feet off the ground and 18 to 30 inches off the wall, well into the airflow. Mount

Figure 8.2. Adjust water flow to the lowest level that will keep the sock moist throughout the schedule.

the water box parallel to the length of the kiln with the temperature sensor 1 inch off the water surface. Any higher and the sock tends to dry out a bit when the fans blow air down on it, giving false high readings. Socks also seem to crust up and harden more. There is no apparent advantage to mounting wetbulbs next to the center coils of double-track kilns. Indeed, due to their nearness to a major heat source, false high wetbulb readings will give you fits with kiln control.

Use ⅜-inch OD copper tubing to run both supply and drain lines to the water box. Smaller lines on the drain quickly clog with sawdust and overflow water onto the floor. Lumber with lots of sawdust on its surface may require you to use ½-inch lines, so keep an eye on sawdust accumulations. When you change socks, make certain the drain hole is clear. Drain lines must slope downward from the box to the end of the tube run. Sags or loops in the drain form airlocks, which prevent draining.

Supply lines need some method for regulating and observing water flow. A shutoff valve to regulate flow is mandatory. There are several sightglass flowmeters on the market that permit you to measure and regulate water flow.

Vent Systems

Most kiln manufacturers mount vents on the roof, although a few place them on the wall. In high-temperature kilns the roof vents form two rows, one on each side of the fan system. One row acts as an exhaust and the other as an intake.

Until a few years ago all vent systems activated pneumatically. This continues to be the predominant method for controlling vents. There are two different pneumatic configurations now in use.

The most common method mounts all air motors and vent linkages on the roof. Air lines lead from the controls to roof-mounted air motors. Each air motor connects by rods, chains, and/or cables to each vent in its row. To open the vents the controller pressurizes the air line and motor diaphragm. As pressure builds the diaphragm presses against a plunger, which in turn pushes against the vent linkage arms. The vents open. To close them the controller simply releases the pressure. Very simple. Yet with simplicity we sometimes accept compromise.

These vent actuators are fine for on/off control. However, for modulated control, where the vents open only as much as required to maintain setpoints, you will find these actuators lacking. I once spent two days

Figure 8.3. Typical vent installation with the vents mounted on the roof.

attempting to adjust the control span on a set of these vents to control by computer. Even with bricks in the manufacturer-supplied counterweight box, the best we achieved was a 2-pound span. The vents began opening at 7 psi and were fully open at 9 psi. However, they did not begin to close until the pressure fell below 3 psi. Simple on/off control was the only real alternative.

On/off vent control does not give good humidity control. To maintain control around a given temperature, the vents constantly cycle open and closed. The wetbulb also cycles above and below the desired temperature. Rather than controlling at setpoint, you actually control within a range of ± 2 to 3 degrees of setpoint. This is not acceptable for high-quality drying.

Precision-modulated pneumatic humidity control requires a positioner to sense actual vent position and adjust the vents accordingly. For example, suppose the controller sends out a signal asking the vents to open to 50 percent. The positioner receives this signal and begins opening the vents, using as much air pressure as necessary, until it senses that the vents are at 50 percent. It then holds that position until conditions change or the controller instructs it to change its opening. Positioners are a little more complicated, yet they are capable of very close humidity control, on the order of ± 1 degree of setpoint.

I have seen one pneumatic actuator capable of good setpoint control without a positioner. These actuators resemble those used on quality steam control valves, only laid on their sides. They are reasonable retrofits for an existing kiln.

A second pneumatic vent configuration installs all vent linkages within the kiln with the actuator mounted on the end wall. Having the linkages within the kiln is a distinct advantage, because all adjustments and most repairs are inside. There is never a reason for the operator to walk on the roof, endangering his safety or damaging roof panels. With all vents linked by heavy pipe shaft rather than rods, vent adjustment is more uniform. These vents open out to the side and moisture-laden air does not have to change direction as it leaves the kiln. Venting is efficient. The installed positioners give exceptional setpoint control.

The primary advantage to all pneumatic vent systems, aside from simplicity, is that they fail closed. This reduces the likelihood of excessive degrade if they malfunction. A few kilns use electronic actuators mounted to some variation of the roof-mounted vent linkage. These actuators require multiple conductor electrical connections. With more moving parts, there is more to go wrong. When they fail, they often fail open rather than closed, as pneumatic systems do. Failing closed extends the schedule without affecting the grade of the lumber to a serious degree, although you may notice more cupping. Failing open shortens drying time, but quality often

Figure 8.4. Typical shaft-driven roof vents. All linkages and linkage adjustments are internal to the kiln. Once this kiln is built, there is little reason to walk on the roof. Photo courtesy Wellons, Inc.

suffers. Since time is paramount in most operations, the shortened drying time is a plus and rarely investigated. Often, many charges dry before grade changes are noticed and the problem corrected.

Adjusting Vents

The vent system must vent moisture evenly. This means that all the vents in a row must open together. In addition, both vent rows must open the same amount when compared to each other. Variations in opening along the kiln length and from side to side cause moisture variations in the lumber.

The standard maximum vent opening is ⅓ of the vent size. For example, if your vents are 18 inches square, they should open 6 inches. Fully modulated vents are not bound to this and may be adjusted higher since the controller will adjust the opening as needed. If you are short on venting area, you may want to increase the opening to its maximum opening. If your direct-fired kiln tends to warp the top layers of bundles, try adjust-

Figure 8.5. Internal view showing the internal linkages for the vents in Figure 8.4. Photo courtesy Wellons, Inc.

ing your openings down a little at a time. Stop when you begin noticing extensions in drying time.

Adjustment Procedure for Roof-mounted Vent Actuators

Tools needed: (1) 6-inch adjustable wrench, (1) 6-foot folding rule, and (2) open-end wrenches to fit linkage bolts.

1. Force the vents open using the control system. Make certain you have full travel on the air motor. Take the air line loose at the air motor. Work the vents open and closed by hand as you check for broken hinges and connections. Make repairs before proceeding.
2. Loosen the linkage between the last vent and the air motor so all vents are down. Now go to each vent and adjust it to a fully closed position. Continue until all vents in the row are adjusted.
3. Reattach the linkage to the air motor and connect the air line. Using the rule, measure the distance from the underside of the vent cover to the top front edge of the vent opening. Take this measurement at the widest point.

4. Adjust the linkage to the air motor until the opening is one-third of the outside square dimension of the vent (a 21×21 inch vent will have an opening of 7 inches).
5. Check each vent in the row to see that the openings are the same.
6. Detach the air line and check that the vents seal when closed. If everything is tight, connect the air line and move over to the second vent row, beginning with step 1 again.

NOTE FOR STEPS 5 AND 6: Deterioration and uneven sagging of roof panels is a common reason for uneven vent openings. Look out for this on flat-roofed kilns. Unless the sagging is corrected, you will have to live with the variation in vent opening.

Adjustment Procedure for Internal Vent Linkages (Shaft)

Tools needed: (2) 8-inch open-end wrenches for adjusting nuts, (1) regular screwdriver, and (1) stick or tape rule.

1. Open vents to full 100 percent position. Inspect them for broken hinges. Make sure you have full rotation on the vent shaft by removing the positioner cover and checking visually. If not, adjust the span screw on the positioner until the shaft fully rotates.
2. Beginning at the end nearest the positioner, adjust the vent to the desired opening based on vent dimension. Measure this distance from the underside of the vent cover to the edge of the vent opening at its widest point. Move down the line until all vent openings are the same.
3. Cycle the vents closed and check for tight closure. Again, be sure that the positioner is rotated fully closed. If not, adjust the zero screw on the positioner until closed.
4. Repeat steps 1 through 3 for the second vent row.

Vent and Wetbulb Maintenance

Check vent linkages at least once a month. With the air signal on, take the air lines loose at the motor and work the vents back and forth. Pay close attention to their connections at the vents and at any intermediate supports. If the linkages must be replaced, use tubular aluminum linkages. They are light, easily installed, and durable. Avoid using cables, chains, or other flexible linkages. They make it difficult to adjust vent openings because there is too much play in them. A few manufacturers use chain to connect the air motor to the first vent. Unless there is a chain attached to the far end of the vent row, the wind will catch these vents and

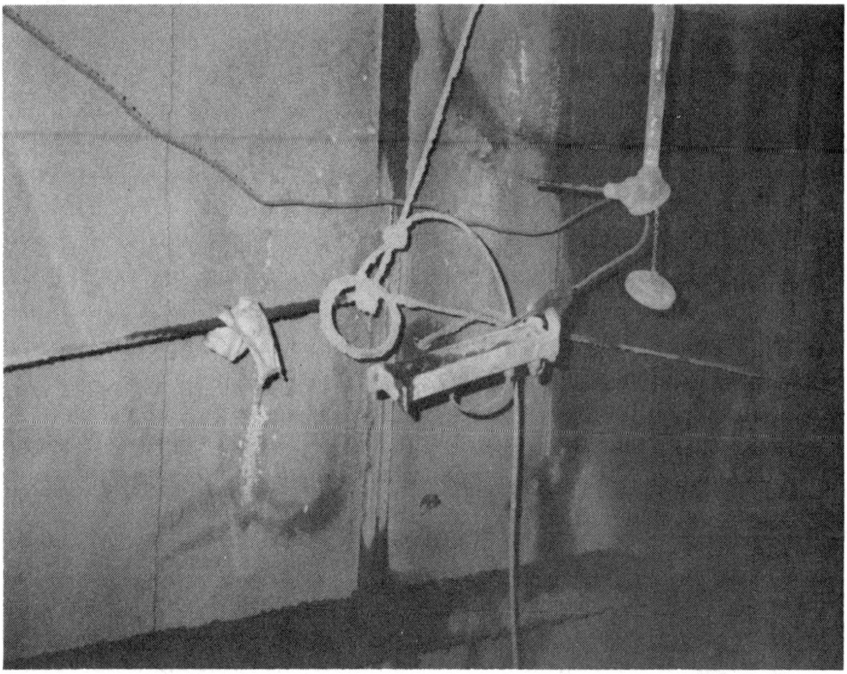

Figure 8.6. Many high-temperature kilns that use their wetbulbs to monitor and not control often fail to keep them in service. They pull the charge on time alone.

blow them open. Being directly in the hot exhaust, these restraining chains rust and fail quickly.

Check the air motor where it attaches to the roof. Air motors with on/off control tend to loosen their bolts and pull out, lifting the motor rather than the vents.

Air motor diaphragm failures are sudden and unpredictable. Keep a new diaphragm on hand at all times and watch your kiln charts for evidence that a failure has occurred. Charts from kilns with one vent row out override wetbulb setpoints in one fan direction. They are almost in control in the other. Air line leaks also appear this way. To isolate the cause, first take the air line loose from the controller. Place your finger over the air line opening on the controller. If the pressure gauge for the air line does not go to full pressure the problem is in the controller. If it does go to full pressure connect the air line and take it loose at the air motor. Place your finger over the end of the air line. If the pressure gauge does not go to full pressure, the air line leaks. If the gauge shows full pressure you have either a bad diaphragm or broken connection at the air motor (ferrules often split).

Use nylon tubing, rather than copper, for air lines. It is available in 250-foot and longer rolls, which permits long runs without intermediate couplings. Copper tubing normally comes in 50-foot rolls requiring couplings for runs over 50 feet. Couplings tend to work loose and leak over time, particularly when allowed to vibrate against the kiln. At 25 percent the cost for copper, nylon is inexpensive to replace as needed. It is also much easier to work with, requiring only a sharp knife to cut to length. It uses conventional ferrule connectors. Nylon is the ideal tubing material for multizone drykilns, with their high requirement for tubing runs. Its service life varies from two to five years, depending on installation. Run it in a separate tray, away from the kiln walls and steam pipes. Some varieties become brittle during extended sunlight exposure. Use covered tray in high-exposure situations and purchase tubing with the highest available UV resistance.

A functioning wetbulb is necessary to control the vents. Change wetbulb wicks no less than every other charge. If they become hard and encrusted by the end of each charge, then you will have to change them then. Encrusted socks give elevated wetbulb readings. Wash new socks before use to remove the fabric sizing left from manufacturing. If you change socks every day, you will save money by washing and reusing them until they show signs of wear. Throw them into a box next to the control room door so you remember to take them home once a week.

Check to see that water is not draining onto the floor. Not only does this upset drying conditions, but the concrete and any steel around it deteriorates. The wetbulb only needs enough water to keep it wet throughout the varying evaporating conditions throughout the charge. Evaporation is heaviest toward the end of the charge, so adjust flow then.

Venting Methods

Until recently, only two venting methods were available to the kiln operator—conventional venting and exhaust-only venting. A third method, demand venting, has become available through the introduction of computer controls.

When the first high-temperature kilns were introduced, they borrowed most of their control methods from the conventional drying systems then in use. Conventional venting opens both vent rows at once when venting is required. In conventional kilns this was the logical approach. The only way to exhaust moisture-laden air is to push it out with drier, cooler air.

Most high-temperature schedules keep the vents closed for a preset timing period to heat the lumber to setpoint before opening. Once the

Figure 8.7. Overflowing wetbulbs pit concrete and degrade structural steel.

timer expires and the vents open, wetbulbs do not control to setpoint but fall freely to a lower control level or as far as they will. Using conventional venting, heat losses are massive and result in a kiln running without drybulb or wetbulb setpoint control. This extends drying time and often causes serious degrade. A few older high-temperature kilns continue to use this venting method. If you have one give serious consideration to converting it to exhaust-only venting.

Exhaust-only venting developed to control the heat loss and degrade attributed to conventional venting. It has become the predominant venting method for high-temperature kilns. It works like this: When venting is required, only the vent row on the exhaust or pressurized side of the fans open. This has several advantages. First, cold air is not deliberately sucked into the kiln. This saves the energy necessary to preheat this air before it becomes useful. Since high-temperature kilns operate above the boiling point, the expansion of water to steam forces the moisture out. This is similar to steam escaping from a tea kettle. Makeup air is not necessary, although a small amount does make its way in through cracks in the walls and around doors. This is the recommended venting method for high-temperature kilns running traditional high-temperature schedules.

Once the drying schedule format for computer controls became flexible enough to accept any schedule for any species, experimentation followed. Schedules were adapted for either venting method. One of the distinct advantages of computerized controls is their ability to ramp setpoints up or down gradually over time. Previously setpoint changes were abrupt, involving fast, radical changes from one setpoint to the next. By combining conventional venting methods (opening both vent rows) with gradual wetbulb setpoint changes, drybulb setpoints were easier to maintain and drying time reduced. Quality also seemed to improve. Still, it was sometimes a little difficult to control with both vent rows opening the same amount.

Enter demand venting. With this method, once the computer determines that venting is required, it begins opening the exhaust vents to bring the wetbulb under control. With the exhaust vents open 100 percent and the wetbulb still out of control, the computer begins opening the intake vents up to their maximum level of 100 percent. The intake vents usually modulate at some level well below this to give very tight control over wetbulb setpoints. This technique prevents over venting, vent cycling, and saves energy.

Traditional Vent Control Methods

Control systems using exhaust-only venting are set up in two ways. Both methods use timers to keep the vents closed during a preset warmup period. Do not attempt to control this warmup period by making setpoint changes by hand. Use a timer. Otherwise, there is simply no charge-to-charge consistency.

The timing period for warmup depends on the item being dried and the individual kiln. With the vents closed the kiln structure and lumber surface comes up to temperature. Two-inch lumber should reach a setpoint of 235° F to 240° F measured in the airstream coming off the coils in 3 to 4 hours. An additional 4 hours is added to get the lumber hot to its center. This helps move moisture to the lumber surface. The total time for the warmup period averages one-third of the total drying time. Determining the actual timing period for your kiln requires trial and error, with grade being the determining factor.

Recent experiments with alternative venting methods indicate that this heat-up period is not mandatory to maintain lumber quality. Indeed, some mills routinely set very large wetbulb depressions from the beginning of the schedule and allow drybulb temperatures to rise slowly. Grade looks good and drying time is compatible with traditional methods. This technique does consume about 30 percent more energy, however.

The difference between the two control methods begins when the vents open or "pop," as operators say. Which method you use depends on the kiln manufacturer. The first method opens the exhaust vents and keeps them open until the end of the charge, allowing the wetbulb to float down. This is often used to determine when to pull the kiln, but results in very harsh drying conditions toward the end of the charge. Without any control over the wetbulb, you don't really know where you are or what you are doing. It is difficult to get consistent quality this way.

A better method also opens the vents after the timer expires, but sets a wetbulb level (from 160° F to 180° F), below which the vents close. As drying progresses the wetbulb often begins to drift below the setpoint. How far it drifts and how fast is used to gauge dryness. The lumber quality from this method tends to be a little better.

Computerized Vent Control Methods

The first computerized controls operated using methods and schedules adapted from traditional drying techniques. Once the full potential of the computer began to be realized, changes followed. The ability of the computer to ramp setpoints up or down has had a significant impact. A warmup period is still required to bring the kiln and lumber to temperature. Yet, by manipulating setpoints, venting may begin earlier. Measurements of how the temperature drops as air passes through the lumber demonstrate that energy consumption falls off during the period between when the kiln reaches setpoint (3 to 4 hours) and venting commences (8 hours). This is, in effect, somewhat idle time. Utilize this timing period better by setting wetbulb setpoints to a level slightly below what they achieve naturally with the vents closed. I usually set 2 to 5 degrees lower than the maximum wetbulb temperature. This releases some moisture, increasing heat demand in what is usually a low-demand period, and speeds drying. This results in a modest reduction in drying time.

When the heatup period has passed, begin reducing setpoints. It is desirable to make a relatively large initial drop of 15° F to 20° F over 2 to 4 hours, and then gradually reduce setpoints to a final level of 140° F to 160° F near the end of the schedule. The quick drop helps control warp with only a moderate – and quickly recovered – heat loss. Low humidities also help keep the lumber bright. At the end of the schedule the kiln runs under both wetbulb and drybulb control to narrow moisture distributions. These are general instructions designed to give you guidelines for finding the right schedule for your kiln. Kilns and controls vary so much in ability and performance that no one schedule fits all.

No Vent Control

A few mills swear that drying without vents is the way to go. Keeping vents closed throughout the cycle means that moisture is released very slowly. This may be an acceptable alternative for kilns that have many leaks in the wall panels or which are direct-fired. This would tend to help get temperatures up. I have even heard of mills building new kilns without installing vents at all. These kilns will have extended drying times because moisture movement is slowed. Non-venting is a response to other kiln problems, be it design or maintenance. Few kilns should have to resort to this.

Using the Wetbulb to Determine Lumber Dryness

Many operators use their wetbulbs to judge when the kiln is ready. To others, the method is unreliable. The level of success you enjoy depends on your specific kiln design and operating method. I use it as simply one indicator, among several, relying on experience to make a best guess.

An operator with tight, well-constructed steam-dry kilns has the best chance of making this method work. Once the wetbulb begins drifting below the final control setpoint, begin keeping close watch over it. For example, if the final setpoint is 160° F, you might find that you can pull the kiln when the wetbulb drops to 156° F. Each kiln is different. Determine the shutdown wetbulb temperature for each kiln individually. Operators whose kilns share common walls or have deteriorating wall panels soon learn that the wetbulb reading is virtually worthless for this purpose.

A good friend of mine says that using two wetbulbs improves the accuracy of this method. He mounts a wetbulb and an extra drybulb on each side of the kiln. As the charge end approaches, George reads both the drybulb and wetbulb from one side (depending on fan direction) and refers to the drybulb/wetbulb charts for the humidity and EMC at that point. Experience has shown him what values he is looking for. He reports that his charge-to-charge moisture variability is significantly reduced, placing 75 percent of his lumber within a 5 percent moisture range. Pretty good for a direct-fired kiln.

Steam Sprays

There has been a renewed interest in the use of steam sprays in high-temperature kilns, primarily for conditioning and equalizing at the end of the charge. The dimension pine markets have become so competitive that more and more mills are looking toward specialty products. To these customers, uniform moisture content relieving drying stresses are very important. These markets require kilns with spray systems.

High-temperature kilns, unless direct-fired, normally require high-pressure steam for heating. Their pressures range from 100 psi to 200 psi, with the largest percentage in the 120 psi to 150 psi range. At least one kiln has been designed to handle 60 psi steam. Even at 60 psi we want to get the pressure down further even though 95 percent of the heat in steam is released as it condenses on the lumber surface.[3] Spraying steam into the kiln in enough quantities to affect humidity often raises the kiln drybulb temperature and we lose control of kiln conditions. Yet somehow we must introduce steam to condition.

The solution is to de-superheat the steam by reducing its pressure while mixing in water to bring its temperature down. The first de-superheater I ever saw was simply a water line stuck into the steam spray system. When the sprays came on water sprayed into the steam flow. There was little control over the quality of the low-pressure steam produced and this system did not work well. We quickly abandoned it.

A good de-superheater uses a pressure-reducing valve (PRV) to maintain pressure in a separate low-pressure steam header. Most operate with a header pressure of 20 psi. As steam enters the header through the PRV, water is sprayed into the flowstream. A separate spray control valve regulates flow from this header to the spray distribution system located just above or below the overhead coils on one side. This distribution system consists of a simple perforated pipe extending the length of the kiln.

Some manufacturers install directional nozzles. Install a drain at the far end of the distribution system and slope the pipe toward it. Traps are unnecessary and do not function if installed on this line.

Direct-fired kilns require a steam generator to supply conditioning steam. The distribution system within the kiln is the same. I recall a mill that installed a perforated, water-filled pipe to supply conditioning moisture. The heat from the kiln supposedly evaporated water from the pipe. Avoid this approach. It does not supply enough moisture to make any impact and is impossible to control. Also, avoid spraying water mist directly into the kiln. This method causes severe corrosion problems.

Spray Control and Scheduling Strategies

Operating the sprays during warmup reduces your time to setpoint by perhaps 15 minutes. However, the excess moisture seems to intensify warp in the lumber by keeping it flexible and allowing it to move out of position. Do not spray during this period.

Toward the end of the charge the sprays can help maintain humidity conditions. By maintaining setpoints they smooth the transition into equalizing and conditioning and help narrow moisture distributions. However, this prevents the use of wetbulbs to indicate dryness.

Kiln operators must face the most difficult task imaginable. Daily they are asked to take green lumber varying in moisture content anywhere from 70 percent to over 130 percent and dry it all within a narrow, 5 percent to 8 percent range. Add to this the variations in drying resulting from nonuniform manufacture, stacking, handling, and loading practices. Finally, toss in the variability inherent in any kiln design, and you begin to wonder how they do as well as they do.

Equalizing

One method for narrowing moisture distributions is running an equalizing schedule at the end of the charge. These schedules work by setting the kiln conditions to an EMC (equilibrium moisture content) that is 3 percent below the desired final moisture content. None of the lumber will fall below this level if the conditions are maintained.

For example, if we want our lumber dried to 12 percent moisture content we would set our wetbulb and drybulb conditions to give us an EMC of 9 percent. Table 8.1 is set up to help you find the equalizing drybulb and wetbulb setpoints by selecting your target moisture content. The 3 percent moisture deduction is already figured.

Table 8.1 ***Setpoints for Equalizing Schedules***

Drybulb setpoints	*Target Moisture Content (% MC)*						
	9	*10*	*11*	*12*	*13*	*14*	*15*
	Wetbulb Setpoints						
150	124	127	130	133	136	138	140
155	128	132	135	138	141	143	145
160	133	137	140	143	146	148	150
165	138	142	145	148	151	153	155
170	143	147	150	154	157	159	160
175	148	152	155	159	161	163	165
180	154	158	161	164	166	169	170
185	159	163	166	169	172	174	175
190	164	168	171	175	177	179	181
195	169	173	177	180	182	185	186
200	174	178	182	185	188	190	191

In Table 8.1, where we desire a target moisture content of 12 percent, we find that a number of drybulb/wetbulb combinations will do the job. Try to equalize at the highest drybulb temperature at which you can control the wetbulb. This allows you to equalize in the shortest period of time. You must determine what is best for your kiln through trial and error. One point is important here: Your kiln must be capable of holding setpoints within very narrow tolerances. Looking at the chart, it is easy to see that for any drybulb setpoint small variations in wetbulb shift you into the next target moisture content.

When do you begin equalizing? The rule for equalizing says to begin when the driest boards reach a moisture content of target minus 3 percent. In the above example we would begin equalizing when the driest board reaches 9 percent. How do we know when the driest board is at 9 percent? We don't, really. No one cuts moisture sample boards for high-temperature kilns. In-kiln moisture meters give you moisture contents of selected boards, none of which are the driest boards in the kiln. Mills running equalizing schedules usually begin them at a preset time, usually a set time before the lumber is expected to be dry. Computerized systems with some method for determining average moisture content and those kilns with internal moisture meters begin equalizing when they reach a certain moisture content. What that moisture content is can only be determined by trial and error.

How long do you equalize? The rule for equalizing says that you run your equalizing schedule until the wettest boards are down to the target moisture content. Again, you must determine what works for you. If you are trying to reduce all your lumber to a narrow moisture distribution, you may end up doubling the time the lumber spends in the kiln. Obviously, some compromise is needed.

When I was a lumber inspector I supervised a mill that practiced its own form of equalizing. The kiln operator simply shut off the heat and closed the vents one hour before the lumber was dry, keeping the fans running. He allowed the kiln to run in this manner for two more hours. This seemed to work well for them and they rarely had moisture problems. The lumber looked good, too.

Conditioning

Conditioning schedules are run at the very end of the charge and serve to relieve drying stresses within the lumber. This is important for lumber that will be remanufactured. Unrelieved stresses cause lumber to warp out of shape and pinch saws as it is ripped or resawn. This is a particular concern for furniture companies. In general, conditioning is not practiced on lumber whose moisture content is above 12 percent.

Conditioning reintroduces moisture into the surface of the lumber. The rule requires that the equilibrium moisture content (EMC) conditions be set at target plus 3 percent. For example, if your target moisture content is 9 percent, set your kiln EMC conditions to 12 percent. Table 8.2 allows you to select the drying conditions for your target moisture content directly. The three percent allowance has been figured in for you.

Again, notice how small variations in wetbulb temperature at any drybulb setpoint affects the target MC. Tight control of kiln conditions is vitally important for successful conditioning. As with equalizing, it is best to condition at the highest temperature your kiln is capable of maintaining. Complicating this is a tendency for the drybulb temperatures to rise as you spray to raise humidity. This is a particular problem at the higher target MCs and drybulb temperatures.

When do you begin conditioning? Start at the end of equalizing or following a cool-down period after your regular charge. Lumber does not condition above 212° F so you must get your temperatures down before starting. Run the fans with the vents open or open the doors to cool the kiln and lumber down. You might even want to push the lumber out for a few hours.

Condition until stresses are relieved. Determine this by cutting samples from the lumber and doing the traditional tine test. In this test a

Table 8.2 ***Setpoints for Conditioning Schedule***

Drybulb setpoints	*Target Moisture Content (% MC)*						
	6	*7*	*8*	*9*	*10*	*11*	*12*
	Wetbulb Setpoints						
150	133	136	138	140	141	143	144
155	138	141	143	145	146	148	149
160	144	146	148	150	152	153	154
165	149	151	153	155	156	158	159
170	154	156	158	160	162	163	164
175	159	161	163	165	166	168	169
180	164	170	169	170	172	173	175
185	169	172	174	175	177	179	180
190	175	177	179	181	182	184	185
195	180	182	184	186	187	189	190
200	185	187	190	191	193	194	195

short, 1- to 1½-inch-long section is removed from a board, about 6 inches from the end. Using a bandsaw, make parallel cuts ¼ inch to ⅜ inch below both board surfaces, stopping about ¾ inch from the end. Remove the center core. If the outsides pinch together, the lumber surface is casehardened and you need to continue conditioning. If they bow apart you have gone too far and reverse casehardened the lumber. There is no remedy for this. If the tines remain straight, conditioning is complete. Congratulations. After a while you will know how long you need to condition and skip using these tests. Keep an eye on grade recovery. There have been reports of lumber that was conditioned and equalized to very narrow moisture distributions, but was heavily warped by the process.

Until recently few pine mills conditioned or equalized their lumber. The reason was simple: It took too much time and effort. Further, it was not necessary for construction-grade dimension lumber, their primary market. Unless a valued customer (with lots of money) has special requirements, mills push it in and push it out in the shortest possible time.

The Australians take a different approach to equalizing and conditioning. The lumber they saw is very prone to warp due to unrelieved drying stresses. Rather than condition within the kiln, they build a separate conditioning chamber for each kiln. Once lumber completes its drying cycle in the kiln it is pushed into the conditioning chamber to finish. This releases the kiln for drying and allows the mill to condition as much as is necessary without impacting on production. This is a pretty good approach.

Today, the pine dimension markets are increasingly competitive. Mills now look toward the specialty markets, where moisture is critical. The next few years will see them develop the special schedules they need to compete in this field. There are no hard and fast solutions that apply to every situation, only guidelines. When you get right down to it, the only right solution is the one that works for you. It is up to you to discover it.

REFERENCES

1. EMC data derived from tables published in the *Kiln Operator's Manual for Eastern Canada,* by M. Y. Cech and F. Pfaff. Ottawa, Ontario: Eastern Forest Products Laboratory, 1977.
2. Equalizing and conditioning guidelines taken from the *Dry Kiln Operator's Manual,* by Rasmussen, E. F., Agricultural Handbook No. 188, Forest Service, GPO, 1961.
3. Hart, C. Arthur. "Kiln Overheat when Conditioning Lumber." *Forest Products Journal* 40, No. 4: 9–14.

CHAPTER 9

Steam-heated Kilns

Progress always involves risk; you can't steal second base and keep your foot on first.

Frederick Wilcox

Steam-heated kilns offer the greatest potential for heat control of all high-temperature drying systems. They have a reputation for producing quality, bright lumber with high-grade recovery and narrow moisture distributions. For export markets, where all these factors are important, it is the drying method of choice.

Steam kilns require boilers to generate steam. This is a significant capital investment. Even so, it does not pay to take shortcuts while selecting the boiler for your operation. You will have to live with the wisdom of your decision for a long while and face it every day. The initial investment in a bargain boiler is quickly overshadowed by mounting operating and maintenance costs. Your boiler must be engineered to perform reliably for many years, but above all it must be of sufficient size to handle your needs. There is nothing more frustrating than trying to run a drying system with inadequate steam to do the job. You never seem to catch up. Worse, inadequate steam means inadequate heat to achieve and maintain control setpoints. This directly affects grade. Mills in this situation are forced to overfire their boilers to maintain production. This shortens the life of any steam generating system. Steam is steam. As a kiln operator I do not care where or how I get my steam, provided it has sufficient pressure and there is enough of it. If I am also responsible for generating my own steam, then that is another matter. I need a boiler that doesn't give me problems. I have enough to worry about just drying lumber.

For mills running high-temperature kilns, running a boiler means running a wood-fired boiler. The high energy demands of these kilns place fossil-fuel boilers out of reach of all but a few. Wood waste is cheap, readily available, and often difficult to dispose of. Using it as fuel solves disposal problems and brings the mill tangible benefits in the form of dried lumber. More mills have also begun investigating the advantages of using a portion of their boiler capacity to generate power. As purchased power costs rise this option becomes increasingly attractive, although not economically feasible for all.

Thermal Oil-heated Kilns

Some kilns heat using thermal oil as a heat-transfer medium. In operation these kilns control and dry similar to steam-heated kilns, although they use pumps instead of control valves and do not require traps. The primary advantage of these systems is the high temperatures they can attain. Steam at 150 psi has a temperature of 365° F. Thermal oil is capable of operating in the 500° F range with only the head pressure of the pumping system. For this reason they are the logical choice for kilns designed to operate using ultra-high temperatures and air velocities (300° F drybulb/2000 fpm air velocity). Sawmills operating adjacent to board plants that use thermal oil for the presses should consider tapping into their heating system.

Thermal oil has seen wide use in Canada due to the extreme risk of freezing in conventional steam-heated systems. Canada is also reported to have relaxed requirements for boiler engineers on these systems. Since they operate at reduced pressures, boiler operators are not required round the clock. Although there is a higher operating cost for electricity to operate the circulating pumps, this may be offset by the elimination of traps and the need to condition boiler water. We will see more of these systems in the future.

Steam Coil

In steam kilns heat transfers into the kiln by controlling the flow of pressurized steam through control valves into heating coils. These coils are fabricated from finpipe. As heat transfers from the steam through the finpipe walls, the steam condenses and drains to the bottom of the coils as liquid condensate or water. This condensate is removed from the coils by traps that separate it from the high-pressure steam. Otherwise, condensate would pool in the coil and no further heat would transfer to the kiln environment.

The amount of steam finpipe you need in your kiln depends on many interrelated factors. Steam pressure, lumber thickness, kiln capacity, and the type and size of finpipe are important considerations. Kiln manufacturers look at all of these and more when designing the heating system for your kiln. Still, coil design is not an exact science. Experience plays a larger role than might readily be admitted. When in doubt it is always best to take the conservative route. Additional finpipe in the initial installation is relatively inexpensive. As long as you do not restrict airflows, you can never have too much. It is much more difficult to add more if you find yourself short.

All high-temperature kilns use finpipe rather than bare pipe for their heat systems. Many finpipe styles are available using different combinations of pipe size, fin thickness, fin height, and fin material. From an economy and long-term service standpoint, schedule 40 steel pipe with heavy rolled steel fin is hard to beat. It is important that the fins tightly contact the pipe for maximum heat transfer. Loose fins are little better than bare pipe. A few manufacturers offer steel pipe with aluminum fins. Although very useful for corrosion resistance in hardwood kilns, finpipe corrosion has not been a recognized problem in high-temperature kilns. These fins offer less resistance to mechanical damage and tend to bend easily when struck. This is a concern on center coils exposed to falling lumber.

Adequate steam pressure is important. Most mills operate their kilns at 100 psi to 175 psi with the large majority of them within the 120 psi to 150 psi range. Again, the heating system must be designed to perform at the available steam pressure. Although successful high-temperature systems have been designed to operate at pressures below 100 psi, this is not generally recommended. At low pressure heat rise is slow, with drybulb setpoints not coming into control until near the end of the charge.

How do you determine if you have adequate heat? Kilns controlling temperatures on the air entering the lumber should reach control setpoints of 235° F to 240° F in less than 4 hours. Kilns controlling setpoints on air after it leaves the lumber should reach a control setpoint of 220° F in 4 to 6 hours.

Coil and Control Zone Design

Double-track kilns are designed with two distinct sets of coils. The main heating coils are located between the two pairs of tracks. Most people refer to these as the center coils or booster coils. A second set of coils, called overhead coils, mounts at or above the overhead fan deck.

The center coils stand upright between the tracks and fill an area roughly corresponding to the area of the kiln load as viewed from the side. At the bottom, they begin at the top of the truck and extend to the top of the load for its full length. The individual coils must run horizontally, with a slight slope toward their drain end to give uniform heat transfer top to bottom. If the coils sag in the middle condensate pools and reduces the heat transfer capability of the coil. It also accelerates corrosion of the coil. Individual center coils must not stand upright. As condensate forms in these vertical pipes, it coats the inside with an insulating film

Figure 9.1. Properly installed center coils run horizontally and have traps for each coil bank. Photo courtesy Wellons, Inc.

that becomes thicker toward the bottom. These coils are naturally cooler at their bottom ends. This contributes to wet lumber in the bottom units in these kilns. Increase center coil heat transfer efficiency by baffling below the coils.

The overhead coils either mount flat or at a slight angle over the outside plenum areas and adjacent to the overhead fan deck. They are mounted with a slight slope for drainage. The traditional method for controlling heat to the overhead coils places both overhead coil banks on the same control valve. As air circulates in either fan direction, it must first pass through one of these heated coils. If the kiln is venting, then preheated air is immediately exhausted, a wasteful loss of energy. Figure 9.6 illustrates how this works. With the introduction of computerized kiln controls, an innovative overhead coil control technique was recently introduced by a West Coast kiln manufacturer. This is known as the vent loss minimizer, or VLM. In this system each side of the overhead coils has its own control valve. With the fans blowing in either direction, air passes through the first coil without preheating and venting occurs without

Figure 9.2. Vertically oriented center coils lose heat transfer efficiency toward the bottom.

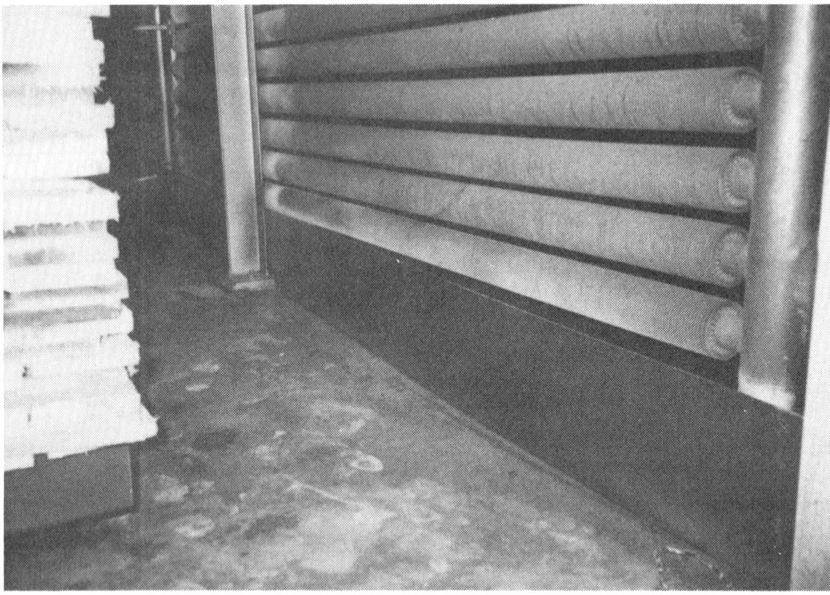

Figure 9.3. If you notice a wet spot in the bottom of your lumber units, baffle beneath the center coil.

Figure 9.4. Overhead coils viewed from the fan deck. Photo courtesy Wellons, Inc.

Figure 9.5. Overhead coils viewed from the plenum chamber with the overhead baffle lowered to show excessive gap between coil and fan deck. This reduces overhead coil heating efficiency.

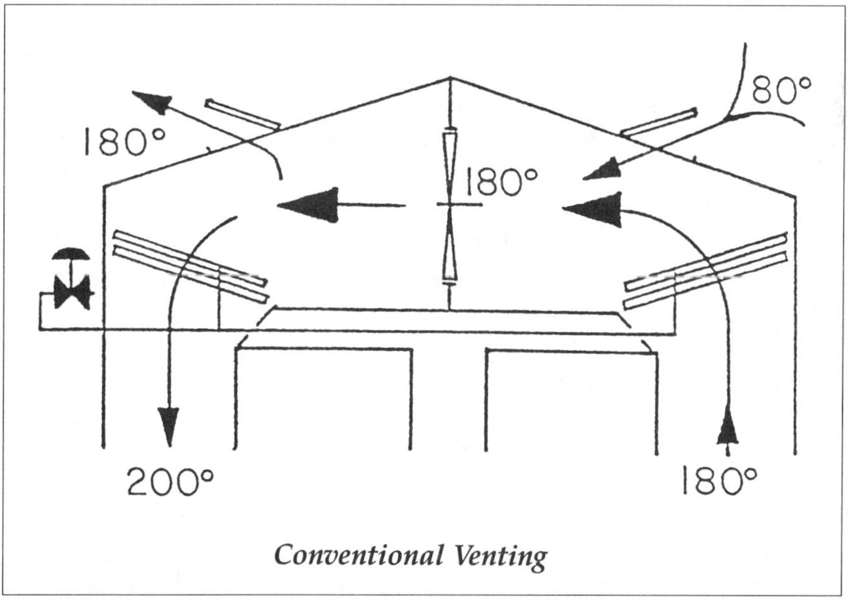

Figure 9.6. Overhead coils with traditional valve configuration illustrating the impact on temperatures while venting.

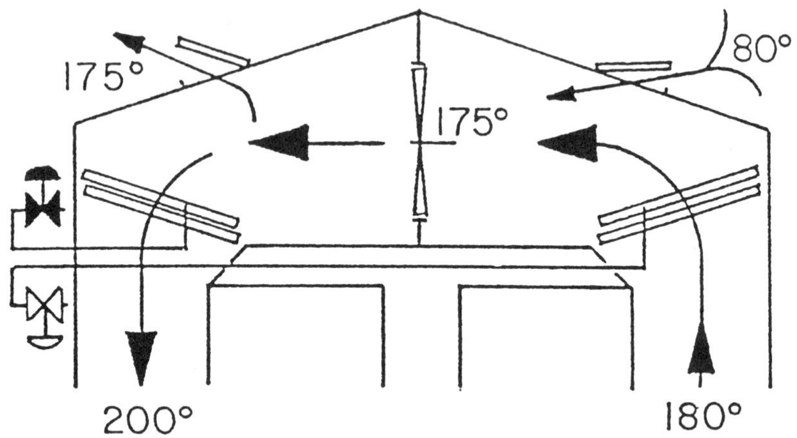

Venting with Vent Loss Minimizer (VLM)

Figure 9.7. Overhead coils configured with the vent loss minimizer (VLM) by placing overhead coils on separate control valves. Less heat is lost when venting. This also illustrates the demand venting process mentioned in Chapter 8.

undue energy loss. As the air continues through the second coil bank it heats as needed to maintain drying conditions. Figure 9.7 demonstrates how VLM works. Adding VLM to the computer system is simple and low cost, requiring only additional valves and software. In addition to energy savings there is also a modest benefit to air velocities. The air that the fans see is cooler and denser, which increases their efficiency. The longer the drying cycle runs, the greater the benefit from VLM.

A few manufacturers mount overhead coils vertically in front of the fans, placing coil banks on both sides of the fans. This restricts access to the fans for maintenance. A few place coils directly at the fan to increase its heating efficiency. Usually there is only one coil bank, which allows them to use less coil and cut costs. There are several drawbacks to this. First, the temperature at the fans is radically different from one fan direction to the next. This regular and extreme thermal cycling may affect the long-term performance of the fan system. It will cause differences in air

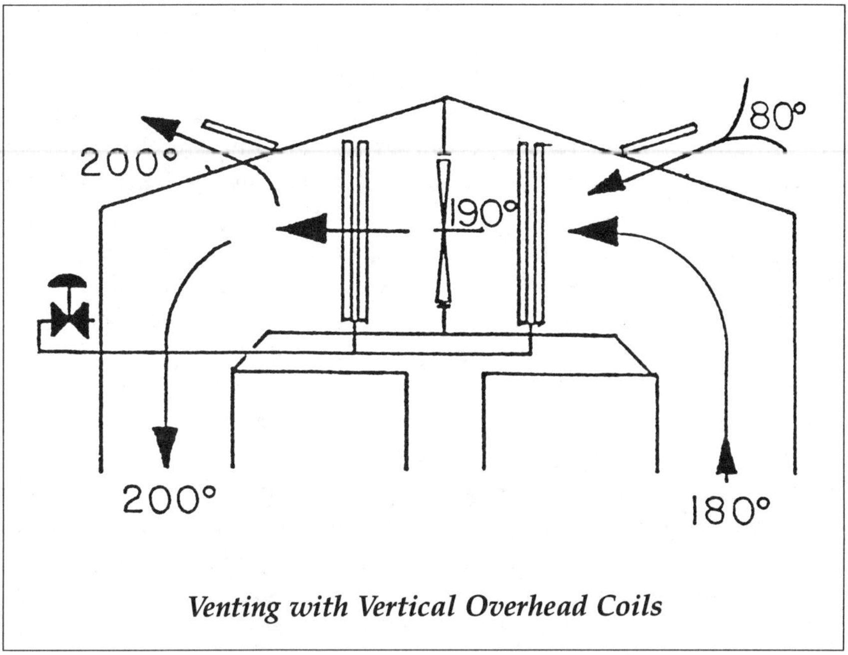

Figure 9.8. Higher heat losses occur with coils positioned next to the fans.

velocity between fan directions. Second, having the overhead heat source between the vents causes excessive vented heat loss because all vented air is completely preheated. It is normal for these kilns to struggle to maintain drybulb setpoints once venting begins. They benefit the most from vent-mounted heat conservation devices, which reclaim exhausted heat. VLM cannot be used. See Figure 9.8.

Multizoning

Until 1982 all high-temperature kilns were built with one or two heat control zones. Single-zoned kilns sense temperatures in one location for each fan direction and modulate steam through one valve to control setpoints. This places both overhead and center coils on the same valve. Since the overhead and center coils are rarely balanced to deliver uniform heat, this zoning technique results in very wide variations in moisture distribution.

Double-zoned kilns are found in several variations. The first divides the kiln into two zones down its length, in effect dividing the kiln into two short single-zoned kilns. Although this helps control lengthwise heat variation to a degree, these kilns are little better than single-zoned kilns in terms of drying quality. The preferred double-zone kiln configuration places the center and overhead coils on separate controls. This method at least permits some control over the heat to each track.

There is one unusual double-zone control configuration I encounter from time to time. This seems to have evolved from the observation that the tops of units are often overdry and warped, while the bottoms are wet. In this approach the center coil is divided in half top to bottom. The top center and overhead coils share a control valve, while the bottom center is on a valve to itself. This causes serious heat imbalances, which actually extend drying time. Imagine a kiln charge coming to setpoint. We know, from multizone kilns, that the overhead and top center coils reach setpoint very early into the charge. These would naturally begin cutting back first. Bottom coils, on the other hand, naturally lag behind the rest of the kiln. By denying the bottom coils the heat contribution of the overhead, they never catch up. Significant improvements in drying are possible by converting these kilns to conventional two-zone control.

In 1980 I conducted moisture and airflow distribution studies in a two-zoned kiln divided into overhead and center control zones. This investigation revealed that as one went top to bottom within the kiln, moisture content increased. Air velocity data showed that air speed decreased top to bottom and demonstrated an inverse relationship between air speed and final moisture content. Figure 9.9 illustrates this air flow and moisture relationship within the test kiln. Several attempts to correct top to bottom moisture variation by baffling the center coils to slow air flow through the top were not successful. Being unable to control air flow effectively, we turned to heat as the control variable.

The test kiln and the proposed kilns were loaded with three units per truck and included two crossout spaces. From the moisture distribution and airflow data we concluded that three heating zones top to bottom with a fourth overhead were required.

Additional data from my studies of an in-kiln moisture monitor had shown wide variations in moisture content down the test kiln's length. Due to the coil configuration of the proposed kiln, we divided the kiln into three longitudinal zones (stations). This gave us two short stations nearest the doors, with a larger middle station. The result was a total of 12 control zones per kiln [(3 center coils + 1 overhead) × 3 stations = 12 zones]. Figure 9.10 illustrates the coil design for these kilns.

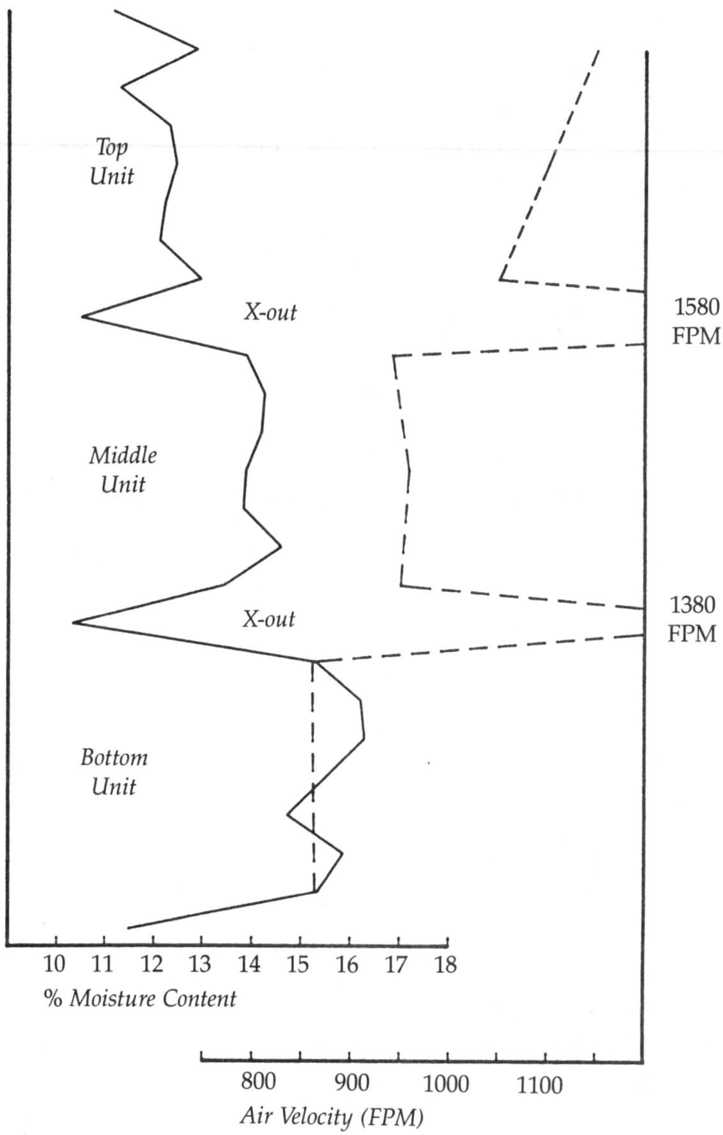

Figure 9.9. Comparison of moisture variation and air velocity top to bottom in one unit of lumber. This data led to the development of the computerized multizone drying system.

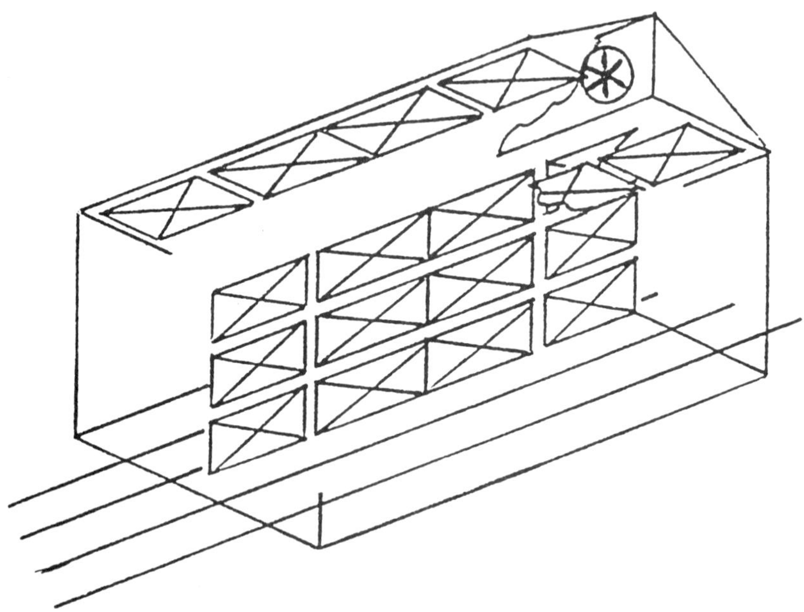

Figure 9.10. Coil configuration in the first computerized multizone drying system. Three zones top to bottom on the center coil with one overhead zone. Three stations down the kiln length for a total of 12 control zones.

The primary benefit of the multizone configuration is the ability to direct heat when and where needed within the kiln to maintain drying conditions. The wisdom of this approach was immediately obvious on the initial system startup. Due to enhanced airflow through the top units, they achieve setpoint far ahead of the lower units. They cut back while lower coils continued to receive heat according to their needs. Conventionally zoned kilns allow the top coils to continue to gain heat as lower coils struggle to setpoint. This contributes to the overdry condition of top units in these kilns. Multizoned kilns do not share this problem. We will discuss the tangible benefits of multizoning in the chapter on kiln controls.

Single-track kilns, because the coils are all overhead, are almost always single- or double-zoned. Recently, a few single-track kilns have been built with multiple zones down their length. This is a distinct advantage, since you are able to control temperatures along the length and reduce variability. Airflow also tends to be more uniform so top to bottom

Figure 9.11. Recent multizone installation with a similar 12-zone coil configuration. Photo courtesy Wellons, Inc.

heat control is considered less important. The primary reason we do not see more single-track dry kilns is that their relative cost per board foot of drying capacity is significantly higher.

Control Valves

Most steam control valves for drykilns are pneumatic. Occasionally I find steam valves with electric positioners, but I do not recommend them. In the event of a failure they may fail open, causing the kiln to overheat. The pneumatic valves found on kilns always fail closed, which adds an element of safety to the operation.

Over the years plug and guided cage valves have proven reliable for drykiln service conditions. These valves give tight shutoff and modulated control of saturated steam. Most kiln manufacturers supply this type as standard equipment. Be wary of other valve types such as butterfly or ball. They do not hold up well under steam control conditions.

Control valves must give tight shutoff against full steam pressure to

Figure 9.12. Multizone pneumatic steam control valves mounted on the catwalk. This system has a shutoff valve on the main steam header, not at each individual valve.

Figure 9.13. Electronic actuated control valves. Control valve rooms are still a good place to keep your lunch warm.

prevent temperatures from overriding setpoint. They must control steam flow throughout their span for accurate temperature control. Never use on/off valves or valves that fail open. I prefer screwed valves from an installation and service standpoint. Flanged valves tend to require more attention to prevent gasket leaks.

Most kiln manufacturers place small shutoff valves ahead of each control valve on the system. The idea is that you can work on any single control valve while the kiln is running. In actual practice this rarely happens. Drying times are so short in high-temperature kilns that most repairs wait until the end of the charge. Safety and comfort is the deciding factor here. I prefer to place one shutoff valve on the main header feeding all the valves to any single drykiln. This allows you to isolate the entire kiln from the steam system for repairs and reduces the number of steam leaks multiple smaller valves produce. It might cost a little more, but the convenience and reduced maintenance are worth it.

A few manufacturers install bypass lines around the control valves. These lines are a convenient way to preheat a kiln before startup. On cold startups they reduce the small steam leaks that often result from fast, uneven heat expansion in the coils. Coils can be heated slowly. In conventional kilns they are used to maintain some steam flow to the kiln while a faulty valve is changed. If the control valve is undersized, opening them during startup increases steam flow. Indeed, this is often their primary function in high-temperature kilns whose valves are undersized.

In actual practice bypass lines are a chronic source of steam leaks and offer small benefit for their trouble. If used to increase steam flow during high demand periods, they are often forgotten and left on. This causes temperatures to override and affects grade.

Steam leaks often occur immediately on cold startup. Reduce them by raising setpoint temperatures slowly, a few degrees at a time. This can also be done by placing small valves in the air lines from the controller to the valves. They are closed and then opened slowly on startup to open the valves and preheat the coils. If your kiln is computerized, simply leave the valves open a little during downtime. This has the advantage of keeping the coils hot and the condensate purged for quick startups.

Weld all possible steam line fittings to reduce leaks. Use socket weld fittings on pipe 2 inches and less in diameter; butt weld on all else. Screwed valves require unions on both sides for valve removal and repair. Standardize valve fittings to make them interchangeable. This allows you to keep a complete valve with fittings on hand and assembled for quick replacement in an emergency. All valves are stamped or cast with an arrow indicating flow direction. Make certain this is correct on new installations and when valves are replaced.

Figure 9.14. Catwalk-mounted pneumatic control valves with an isolation valve on the supply line ahead of the valve.

Figure 9.15. Pneumatic control valve mounted with shutoff and bypass valves. This is a flanged valve.

As with the vents, I prefer nylon tubing for air lines to the valves. This is less expensive than copper and permits very long runs without couplings, which are prone to leak. Run it in covered tray for full support and protection.

Control Valve Maintenance

If your heat tends to override setpoint toward the end of the charge, your valves are bleeding through. On older systems this is an indication of seat wear and requires valve replacement or rebuilding for correction. In newer systems the valve may simply not be seating. To adjust, listen to the valve using mechanics' stethoscope touched to the valve body on the outfeed side. These stethoscopes are available from most auto parts stores.

A bleeding valve makes a hissing sound. Using a large open-end or channellock wrench, turn the bolt on the valve stem to seat the valve. You should see the stem move down as hissing slows and stops. If it does not then (1) there may be trash in the throat of the valve; (2) the valve seat is worn and must be repaired; or (3) the valve will not close completely against full steam header pressure.

Keep extra valve diaphragms on hand in case of failure. As with the vents, diaphragm failures are abrupt and unpredictable and appear as loss of kiln temperature. Temperature loss is also caused by trap failure and the loss of steam pressure. To determine which, first make sure you have full steam pressure. Next, look at the valve. If the valve is fully open suspect the trap or condensate system. If it is closed disconnect the valve air line from the controller and place your finger over the controller outlet. If the gauge for that air line does not go to full pressure, then the problem is in the controller. If it does go to full pressure, connect the air line and disconnect it at the valve. Place your finger over the air line. If the gauge does not go to full pressure, the problem is in the air line. If it does go full pressure, the problem is in the diaphragm or in the air line connection to the valve.

Steam Traps

Inverted bucket traps are the standard trap for high-temperature kilns. Here is what *Heating/Piping/Air Conditioning* had to say in an article about traps:

> The inverted bucket trap provides a combination of advantages that are not available in any of the other trap models. The operation of the trap is intermittent and provides for excellent [condensate removal] action. The

Figure 9.16. Inverted bucket traps mounted in the kiln. Note the isolation and test valves. These traps are easy to test and repair.

> inverted bucket also offers excellent resistance to wear, corrosion, dirt and hydraulic shock. It continues to operate at full capacity on very light loads and against high back pressure. Finally, the inverted bucket trap has established its long-term ability to conserve energy for a much longer period of time than any other steam trap available today.

High-temperature kilns operate under widely varying condensate loads. Being a batch process, the entire steam system suffers wide temperature swings. Dirt and slag in the condensate is often a problem, particularly on new startups and in older systems. Poorly designed steam distribution systems suffer damaging hydraulic shock or "water hammer," the loud "thunking" one often hears before, and immediately after, startup. Water hammer results from poor drainage and can cause severe damage. No other trap better serves this wide range of service conditions.

Never install orifice traps in a high-temperature kiln. These traps are simply holes in a plate welded in the condensate line. The hole diameter is determined by the average steam load of the system and is designed to pass condensate, or steam, continuously. The average steam demand is a far cry from the extremes in condensate load found between startup and shutdown in these kilns. They remove condensate too slowly at the beginning of the charge and waste steam at the end. They are a bad idea.

Inverted bucket traps do not handle air in the system very well. For this reason many manufacturers install thermostatic air relief valves on their coils. Although I realize the importance of removing this air from the system, I have always found these air vents gave problems. Being nonessential to the operation on the kiln (from the kiln operators' point of view), they are seldom serviced. When they fail they blow live steam into the kiln, upsetting kiln conditions. Upon first opening a kiln, while making that first walkthrough to pull back the baffles, I am listening for steam leaks. These vents seem to whistle and blow more than they should and I always end up plugging them to eliminate false alarms. I have never observed a difference in kiln performance. Still, the decision is up to you. To avoid any controversy over this issue, leave a little steam on the lines to keep the coils hot between charges.

I like to install my traps in the kiln. This keeps them at a convenient working level (as opposed to traps in a pit), and reduces their chances for freezing in the winter. With kiln charges so short, repairs wait until charge end. Check traps between charges by simply turning on the heat and listening to each trap in turn. Do this once a week as part of your preventive maintenance program. If you have a computerized, multizone drykiln, check them every day by comparing entering and exiting temperatures through the coils. No heat rise through a coil means you have a problem, not necessarily related to the trap. Traps outside in a pit are difficult to work on. Poorly maintained traps are impossible to check. You might as well have them in the kiln.

Install traps as far below the coil as possible for drainage. Again, weld all possible joints and standardize your trap assemblies. Keep spare traps already assembled with the required nipples and unions for quick change-out. Make certain each trap has a strainer to remove trash and put a valve on it for quick blowdown and drainage. Never place check valves in the condensate line. You can never be certain they work. Make sure the flow arrow on the trap points toward the drain.

Most kiln manufacturers install shutoff valves before and after their traps. There are two schools of thought here. With shutoff valves you are able to quickly isolate and replace traps. Valves are also used in one method for checking trap operation. This is discussed in detail later. With so many additional valves it is important that this system be well designed and maintained. The second method places no valves at the traps but one in the main condensate line to isolate the system. To change traps the main steam and condensate line valves are closed. Traps are checked by

Figure 9.17. Traps mounted in an outside pit. Steam leaks usually render these traps impossible to service without shutting down the kiln.

Figure 9.18. Inverted bucket trap with two isolation valves and unions, but no test valve.

listening to them. The primary advantage of this system is the reduction of potential sites for steam leaks.

At least one kiln manufacturer installs a drain leg in the condensate line leading to in-kiln traps. This drain leg extends out the kiln wall, where a valve is attached. Its purpose is to blow down excess condensate from the coils during startup. This is wasteful and serves no practical purpose. Remove these lines (and the repair headaches) and send the condensate back to the boiler.

Each coil needs its own trap. Tying two or more coils together on one trap contributes to cold spots within the kiln. As you move farther away from the trap, condensate levels tend to rise in the coils served by that trap. The only exception is double-banked coils set side by side. One trap will serve each double bank with no problem.

Checking Trap Operation

Check your traps for proper operation at least once a week. If you have a computerized multizone kiln control system with temperature sen-

Figure 9.19. Inverted bucket trap without isolation valves but fitted with a blow-down line for testing and purging condensate at startup.

sors at each coil bank, use it to check them every day. Doublecheck manually once a week.

To check traps through the computer system, examine the display screens or printouts that give you information on temperature and valve openings. It is best to do this during startup when the valves are 100 percent open and the kiln has been running at least an hour. Look closely at the temperatures on either side of each center and overhead coil. If the air temperature does not rise as it passes through a coil then you can suspect a problem.

First, verify that the computer is telling the control valve to be open and that it is indeed open. If not, determine the reason. If the control valve should be open and is not, take the control valve air line loose from the computer output device (I/P transducer) and place your finger over its outlet. If the pressure gauge does not go to full pressure, then either the I/P transducer or the computer electrical output is faulty. Determine which by using a multimeter to check for an electrical signal from the computer. If you have a signal, then the transducer is bad. If not, then the computer output module is suspect. Look for a blown fuse on the computer output. If the gauge went to full pressure connect the air line and take it loose from the valve. Place your finger over the air line end. If the gauge shows full pressure, then the valve diaphragm or air line connection is faulty. If it does not you have a leak in the air line.

If the control valve is open, then the trap is suspect. If it is early in the charge and the trap is accessible, shut the kiln down and test it manually. Otherwise, complete the charge and mark the lumber in the vicinity of the coils served by the bad trap for special handling. There is a strong likelihood that it is wet.

There are two recommended methods for checking traps depending on how your traps are set up. They are the visual method and the audible method.

Visual Method

This method is used on traps installed with valves on either side and a smaller test valve on the top of the trap.

1. With the heat on, close the valve on the outlet side of the trap.
2. Open the test valve on the top of the trap. Be certain that its outlet points away from you or any bystanders.
3. If the trap is working condensate blows outward, starting and stopping at regular intervals.

Figure 9.20. Testing a trap by spraying water on it. Water boils violently on a working trap. Unfortunately, it does the same on one which is blowing through.

4. If steam blows out continuously check the following:
   - The trap may have lost its prime. Close the inlet valve and let the trap sit a few moments to let condensate accumulate. Open the inlet and test again.
   - The valve may not be seating properly due to scale lodged in the orifice or worn parts. Pull the trap down and replace the kit if needed.

   NOTE: Occasionally I find a blowing trap that goes to working when struck with a stick or other object. Keep your eye on it. If it happens again, replace the kit.

5. If condensate blows continuously, then the coils are probably waterlogged. The trap will be cold. Check the following:
   - The steam pressure may be too high because the wrong pressure was specified originally.
   - The pressure was raised without installing a smaller orifice.
   - The pressure relief valve has failed.
   - The boiler pressure gauge reads low.
   - Normal wear has enlarged the trap orifice.
   - A high vacuum pressure in the return line has increased the pressure differential beyond the trap's capability.

6. If no condensate or steam reaches the trap, consider the following:
   - The strainer ahead of the trap or the pipe line or elbows may be plugged.
   - A valve in line to the trap may be broken.
   - The trap mechanism may be worn or defective.
   - The trap body or vent may be filled with dirt.

7. If the trap is hot but no condensate is discharged:
   - The trap may be installed above a leaky bypass valve.
   - There may be a leak in the blowdown valve in the condensate line feeding the trap.

Audible Method

This method is useful on any trap installation.

1. Turn the heat on to the kiln. Using a mechanic's stethoscope available from an auto parts store, touch the metal probe end to the top of the trap near its outlet.
2. Listen to the sound:
   - A properly functioning trap makes a loud "breathing noise" as the trap fills and discharges. Often the clattering sound of the inverted bucket is heard as it drops after discharge.

Figure 9.21. Using a mechanic's stethoscope to listen to a trap. Do this once a week. A functioning inverted bucket trap sounds like heavy breathing.

- A continuous rushing sound has two causes. Splash a little water on the top of the trap. If it boils violently, then the trap is blowing through. The trap must be repaired. If the water sits without boiling, then the trap is still discharging cold condensate from the system. Allow time for the lines to clear and test again.

- No sound indicates that the trap is dead. Check cutoff valves to make certain condensate may pass freely. Repair the trap if no restriction or blockage is found.

Establish a preventive maintenance program for your steam traps. Inspect each trap and replace its kit once a year. With a little thought, you need only rebuild a few traps each month, rotating through all of them by the end of the year. PM programs head off trap problems and allow you to operate at maximum efficiency and energy savings.

Steam System Maintenance

Finding and repairing steam leaks is often an ongoing project at the drykiln. Not only do steam leaks introduce uncontrolled moisture into the kiln, they cost you money for fuel and water treatment. Mills with chronic, uncontrolled steam leaks often find that the boiler runs easier after replacing deteriorating coils and repairing leaks. The following table illustrates the money even small steam leaks represent. Numerous small leaks easily add up.

Figure 9.22. Repair major steam leaks when discovered. This water erodes concrete, corrodes structural steel, and upsets drying conditions.

Table 9.1 ***Cost of Steam Leaks (3)***
Dollars per year at 150 PSIG

		Cost of steam leak, when steam generation costs:		
Hole diameter in inches	*Leak rate lb/year*	*$4 per 1000 lbs.*	*$6 per 1000 lbs.*	*$8 per 1000 lbs.*
1/16	160,000	$ 640	$ 960	$ 1,280
1/8	635,000	$ 2,540	$ 3,810	$ 5,080
1/4	2,500,000	$ 10,000	$ 15,000	$ 20,000
1/2	10,200,000	$ 40,800	$ 61,200	$ 81,600

Check your kiln at least once a week for steam leaks. In addition, when you change-out charges be on the lookout for water on the floor and rust marks on coils. Dark streaks on lumber next to the center coils also indicate steam leaks. When you first enter your kiln at the end of a charge, listen for leaks and investigate all you hear. Leaks in drain and condensate lines spray water and are easy to spot. Leaks in feed lines only make noise. You cannot see the steam escaping from them in the hot kiln. These leaks must be tracked by sound. I was once so close to a large leak that the noise was unbearable, yet I could not see the leak. Only after I passed my hand quickly over it was I able to pinpoint it. Be careful when doing this! If it is not convenient to repair small leaks when found, mark them for later attention. Large leaks demand immediate action.

REFERENCES

1. Armstrong Machine Works. *Steam Conservation Guidelines for Condensate Drainage.* Three Rivers, MI: Armstrong Machine Works, 1976.
2. Culpepper, L. G. "Application of Computerized Controls to Improve the Lumber Drying Operation." Seattle, WA: IUFRO Meeting, 1989.
3. Hammond, H. "Principles of Operation to Be Considered in Selecting Steam Traps." *Plant Engineering* (July 1985): 70.
4. Makansi, J. *Managing Steam.* New Jersey: Leslie Company, 1985.
5. O'Dell, L. R. "Steam Traps—Basic Facts and Energy Saving Tips." *Heating/Piping/Air Conditioning* (May 1977).

CHAPTER 10

Direct-fired Drykilns

We are continually faced by great opportunities brilliantly disguised as insoluble problems.

Author Unknown

Method of Operation

Direct-fired drykilns are heated by blowing hot combustion gases directly into the kiln environment. These gases are generated by burning oil, gas, or wood waste. High fossil fuel costs have resulted in the conversion of most direct-fired high-temperature kilns to wood waste fuel. The burners for these kilns burn dry, hammermilled planer shavings or green sawdust.

Shavings burners require a small, uniform particle size for efficient combustion. Shavings are blown directly from the planermill into a hammermill, which sizes the fuel and blows or conveys it to storage bins. This fuel is metered into the burner, where it mixes in a cyclonic action with air as it burns. The products of combustion (heat, water, carbon dioxide, and ash) enter an air circulation system for distribution into the kiln. These kilns are characterized by the ash, which collects on the lumber and imparts a grey appearance. The ash is completely removed during planing. These kilns are generally not suitable for mills marketing rough dry lumber.

Sawdust burners are recent introductions that burn green sawdust directly on a grate, not unlike a boiler firebox. Since little dust becomes suspended in the airstream, these burners generate less ash. Lumber from these kilns is somewhat brighter, although this varies between manufacturers and the individual level of maintenance. If overfired, they can also produce ash. For the purpose of this discussion, we will consider all burner types the same.

Direct-fired kilns are somewhat less expensive to purchase than steam kilns, although the more kilns you plan to build, the more competitive steam becomes. However, with reduced capital costs, there are trade-offs.

Figure 10.1. Outside view of a direct-fired kiln prior to vents opening.

1. Ash buildup on the lumber is a marketing problem for many. Beyond this, its presence in the workplace is coming under increasing scrutiny by the federal government. This promises to become an important issue for these burners in coming years.
2. Grade recovery runs somewhat below what is possible in a steam kiln. There is no doubt that a well designed and maintained direct-fired kiln is capable of approaching the grade recovery in a steam kiln. However, proper operation is very important to their success and is a significant factor in the economics of their continued use.
3. These kilns are prone to fires, which occur most often toward the end of the charge, when the lumber is driest. While checking airflows in these kilns, it is common to observe sparks swirling about. Maintenance and burner style have a tremendous impact on this danger.
4. These kilns require more electrical horsepower for their operation than a comparably sized steam kiln, including boiler. Often the circulation blower horsepower alone equals the entire boiler.

Figure 10.2. Suspension burner designed to burn hammermilled shavings. Photo courtesy of McConnell Industries.

Quality considerations in direct-fired kilns are more strongly related to the efficiency of the heat distribution system than anything else. This is because direct-fired kilns have potential problems peculiar to their design. These can be overcome to a degree. Balancing heat delivery height-wise and along the length helps, but total improvements are limited by the single-zone control capabilities of these kilns. A few double-zoned direct-fired kilns have been built. However, these kilns were zoned by sectioning the kilns lengthwise. In effect combining several single-zoned kilns. This does not have any impact on the top to bottom variations in moisture produced by these kilns and addressed so well in multi-zoned steam kilns.

Keep direct-fired kilns short. Balancing heat and airflow becomes more difficult with increasing kiln length.

Heat Distribution Systems

Direct-fired kilns have a ductwork system that distributes heated air from the burner into the kiln. From the burner heated air enters a blend chamber where it is mixed with cooler air returning from the kiln through

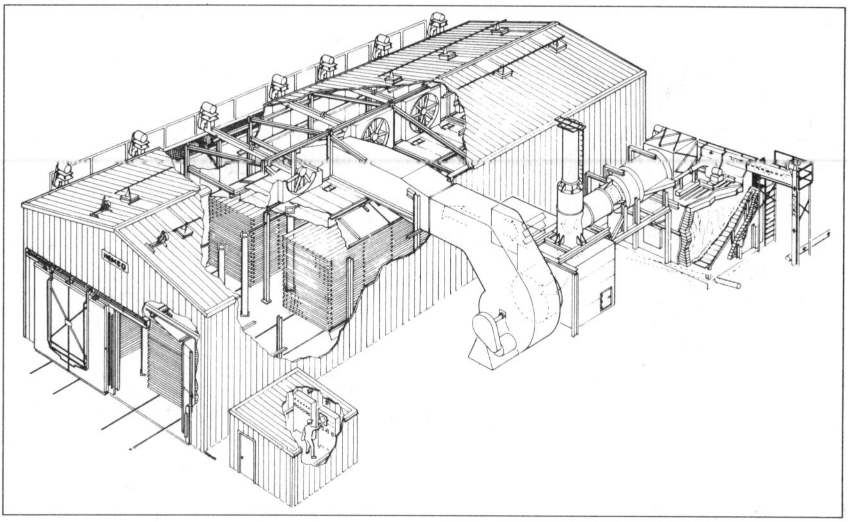

Figure 10.3. Schematic drawing of a direct-fired kiln equipped with a burner designed to burn wet sawdust. Illustrates burner, blend chamber, return air duct, recirculation blower, supply and distribution duct, and downcomers. Drawing courtesy of HEMCO.

the return air duct. From the blend chamber air is moved to a main distribution duct running the length of the kiln. A large centrifugal fan pulls air from the kiln and through the blend chamber, blowing it into the distribution duct. Some manufacturers place this fan ahead of the blend chamber, forcing the burner to blow against its output pressure. Both the distribution and return air systems are sources of imbalances in heat and airflow.

The main distribution duct may be underground in older systems, although this design is not seen very often today. Heat losses through the ground are too high and distribution efficiency poor. The smaller ducts branching up from this are called risers, since they rise from the ground to the upper fan deck area between the two sets of tracks.

Modern direct-fired kilns place the main distribution duct overhead as an integral part of the fan deck. This allows delivery of heat to a central location for uniform distribution. Avoid any design that places this duct on one side of the kiln. From this distribution duct, heated air flows in two directions. Downcomers extend down to the ground between the two sets of tracks and serve as the main heat distribution path. They are square or diamond shaped in cross section, slotted on two sides along their length, and fitted with dampers on their inlets. Additional openings are cut into the top of the main duct and fitted with slide gates or other airflow adjusting mechanisms. This introduces air into the overhead fan deck area.

Heat Balancing a Direct-fired Kiln

The real trick to balancing moisture distributions is balancing heat delivery. This is a difficult and time-consuming process. It involves balancing heat delivery from the downcomers along the kiln length. We then balance the downcomers with the overhead duct openings. Success is often illusive, since any adjustment to any part changes the balance of the entire kiln. Close a damper or slide gate here, and the air pushes out somewhere else. This is truly an art.

Before beginning, wash the kiln to remove ash buildup. Pay close attention to the slots in the downcomers – they must be completely free of ash. Ash buildup is a serious and continuing problem in direct-fired kilns. It gradually accumulates, restricting and rerouting airflow through the distribution system, upsetting heat balance. Indeed, you may find it necessary to wash frequently through balancing procedures lasting more than a few charges.

Downcomer Washing Procedure

Using a firehose, direct the water stream into the downcomer slots from top to bottom. Make certain all ash is removed from both slots in each downcomer. Wash all the downcomers in the line. Since you can wash only one side of the slot from any given track, switch to the other track and wash from the other side. Only then will you remove all of the ash accumulation. Make cleaning a regular part of your maintenance schedule.

Inspect the downcomers for damage. Look for broken or loose nuts and spacers. Check the downcomer dampers. These dampers are easier to adjust if they have handles on them. Installing handles aligned with the damper blades makes it easier to check damper position from the ground. Make certain the dampers have locknuts to keep them in position.

Turn on the recirculation and combustion air blowers. Inspect the downcomers again. Also inspect above and below the overhead distribution duct. Bolts often fall out of the duct and are unobserved until the duct is pressurized. Repair any gaps that allow the uncontrolled escape of heated air into the kiln.

You are now ready to begin balancing the downcomers.

Downcomer Balancing Procedure

Tools required: Screwdriver or Allen wrench to loosen damper setscrews; channellock wrench if the dampers do not have handles; lumber crayon; and an airflow meter.

Figure 10.4. Cleaning downcomers prior to heat balancing. Ash buildup seriously affects heat distribution in these kilns.

1. With the recirculation and combustion air blowers on, begin measuring the airflow out of each downcomer slot at chest height. Write these readings on the downcomer. The average of both

Figure 10.5. Balancing data written on the downcomer for reference. This damper is equipped with a damper just above the markings and has another at the top where it connects to the distribution duct.

downcomer slots is the value for that downcomer. Average the flow from all of the downcomers and use this value to adjust flow.

2. Begin adjusting dampers to even air delivery. It is best to have an assistant turn the dampers while you watch the airflow meter. Make a first pass down the dampers adjusting the airflow within range. A second pass is required to fine tune velocities. Table 10.1 shows sample downcomer readings.

Downcomers nearest the connection between the supply duct and the distribution duct usually have the highest velocities (numbers 5 and 6). Closing these pushes additional air out toward the ends. There will always be one or two downcomers that never seem to get enough air, even when fully opened (number 7). This is caused by air rushing past the top of their openings, unable to make the downward turn. Correct this by installing 3 x 3 angle-iron scoops within the main distribution duct to

Table 10.1 ***Sample Downcomer Readings (feet per minute)***

Downcomer Number	Slot	Before adjustment Velocity Reading	Before adjustment Velocity Average	After adjustment Velocity Reading	After adjustment Velocity Average	Change
1	1	1240	1320	1670	1685	+365
	2	1400		1700		
2	1	1670	1740	1670	1670	−70
	2	1810		1670		
3	1	1260	1315	1660	1655	+340
	2	1370		1650		
4	1	1300	1330	1680	1660	+330
	2	1330		1660		
5	1	2320	2310	1740	1690	−620
	2	2300		1640		
6	1	2270	2270	1660	1650	−620
	2	2270		1640		
7	1	840	935	1660	1645	+710
	2	1030		1630		
8	1	2020	2165	1690	1675	−490
	2	2310		1660		
9	1	1000	1080	1690	1660	+580
	2	1160		1630		
		Average	1607		1666	
		Maximum	2310		1690	
		Minimum	935		1645	
		Range	1375		45	

direct this air into the deficient downcomer. I agree, this is not easy. The only reason I even attempted this was that my Daddy was short. It works, though, and is the only reason I was able to balance this kiln so close.

Do not discount the difficulty of achieving good balance. Although it appears that I balanced this kiln in one simple step, there were actually three intermediate balancing passes and readings made in this kiln. The entire process consumed an entire day.

At one time I believed it was important that the downcomer slots be tapered from top to bottom, with the slot widest at the bottom. I reasoned that the air pressure must be highest at the top of the downcomer and more air must escape here. This would explain the warped lumber at the tops of units. The next time you check the inside of your kiln with the circulation blower on, take a piece of ribbon and hold it in the airflow from a downcomer slot. Move it up and down along its length and observe how the ribbon follows the flow. You will discover, as I did, that air does not blow straight out from these openings, perpendicular to the length of the duct. It tends to flow downward toward the floor. Tapering the slot is less important.

Since there are so many openings in the fan deck level, balancing these is a real challenge. Not only are these openings affected by the openings in the downcomers below them, fan direction also influences how much air is drawn from any one opening. Also, anything you do here affects airflow to the downcomers. Although I have known individuals who claimed they could adjust these by sight, mere mortals must rely on trial and error. You make an adjustment, run a charge, make another adjustment, and so on. How do you know how well you are doing? Most direct-fired kilns are installed with high-temperature limit sensors in the roof. These sensors mount on both sides of the fans at regular intervals along the kiln length. Although a little hard to read with accuracy, these sensors often supply the only information available to aid balancing the overhead fan deck openings. These readings are not reliable for balancing the kiln unless your kiln is fully loaded and properly baffled. Openings that permit circumvention of air around the lumber cause hot spots and erroneous readings.

Keep in mind that higher-pressure areas deliver more air through a given opening. As such, openings tend to be smallest near the supply duct junction and open gradually toward each end of the kiln, with one exception. At the supply duct junction they need to be open a bit more to overcome the reduced openings in that area. Simple, right? Now you have an appreciation for the art of balancing these kilns.

Checking your progress in any heat balancing attempt is much easier with multiple heat sensors positioned at strategic locations within the kiln. Permanent thermocouple networks designed by the kiln manufacturers or in-house are the best. These systems allow you to monitor the kiln for changes that will occur over time. They are also easy to use. A good system prints out temperature data at regular intervals and presents it in an easy-to-understand form. It is frustrating and time consuming to have to pull data out of column after column of numbers to make any sense of it. Also, the easier it is to use, the greater the likelihood it will be used on a continuing basis. One note of caution: If your sensor network uses two wire RTDs instead of thermocouples for temperature sensors, be certain they are in calibration. I once spent a week trying to tune a hotspot out of a kiln only to discover that the hotspot was a sensor 20° F out of calibration.

For temporary measurement rent a multipoint recorder from an instrument house. Inexpensive 22-gauge thermocouple wire with nylon or polyurethane insulation is available off the shelf and may be discarded after use. For extended testing, or for use in several kilns, consider teflon-insulated wire. You need sufficient wire to make four separate runs of sensors down the length of the kiln. Place a sensor on both sides of the space between each pair of downcomers. This allows you to read entering and leaving temperatures from the downcomers. Run the same spacing of sensors in both plenums to measure heat from the overhead area. Make your thermocouples up by attaching one end of the wire to the recorder, pulling the wire into the kiln, and clipping it to length. Strip the leads back about 1 inch and twist them together. Slip a crimp connection over it and crimp. Trim any exposed ends. Dip the end in RTV silicone sealant and staple the thermocouple to the lumber, being careful that it is in the airstream. For ease of handling between charges, tape the leads in each run into bundles.

Use your thermocouple network to determine where your hotspots are, making adjustments to dampers between charges. Over time you will narrow the heat distributions in the kiln. This reduces moisture variability in the lumber. Do not allow ash buildup to fool you into making erroneous changes.

Balancing Return Air

So far we have discussed kiln balancing in terms of balancing heat delivery. If you run a few charges with a thermocouple network, you will soon discover that hot and cold spots seem to move about the kiln with fan reversals. A contributing factor in this is the action of the return air duct.

The return air duct is necessary to return kiln air to the blend chamber for reheating. Almost all of the air distributed by the recirculation blower is returned through this duct. The difference is in the additional air added to the system by the combustion air blower. That air volume escapes the kiln through the vents and cracks in its walls. Still, a tremendous volume of air is recirculated.

Most kiln designs use a single return duct mounted off-center on one long wall. Since virtually all kiln air is migrating toward this duct, natural airflow patterns are radically altered. The duct has a boosting effect on air velocity when air circulates through the lumber toward it. It reduces air velocity in its vicinity when fans reverse. Figures 10.6 and 10.7 illustrate the variations in air velocity in these kilns by using velocity profiles. A computer produces these images by assigning a symbol to air velocities

Figure 10.6. Air velocity profile in a direct-fired kiln with the air blowing toward a single return air duct. Duct location drawn in. Average velocity is 940 fpm with a velocity range of 635 fpm.

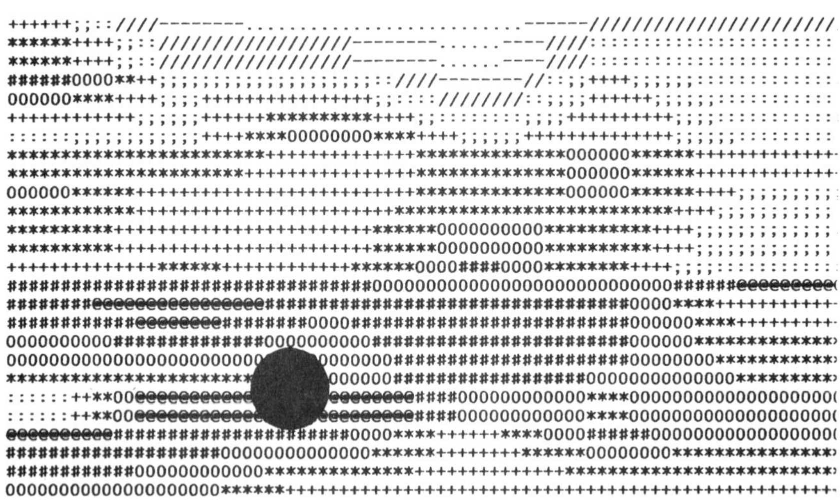

Figure 10.7. Air velocity profile in a direct-fired kiln with the air blowing away from a single return air duct. Duct location drawn in. Average velocity is 823 fpm with a velocity range of 600 fpm.

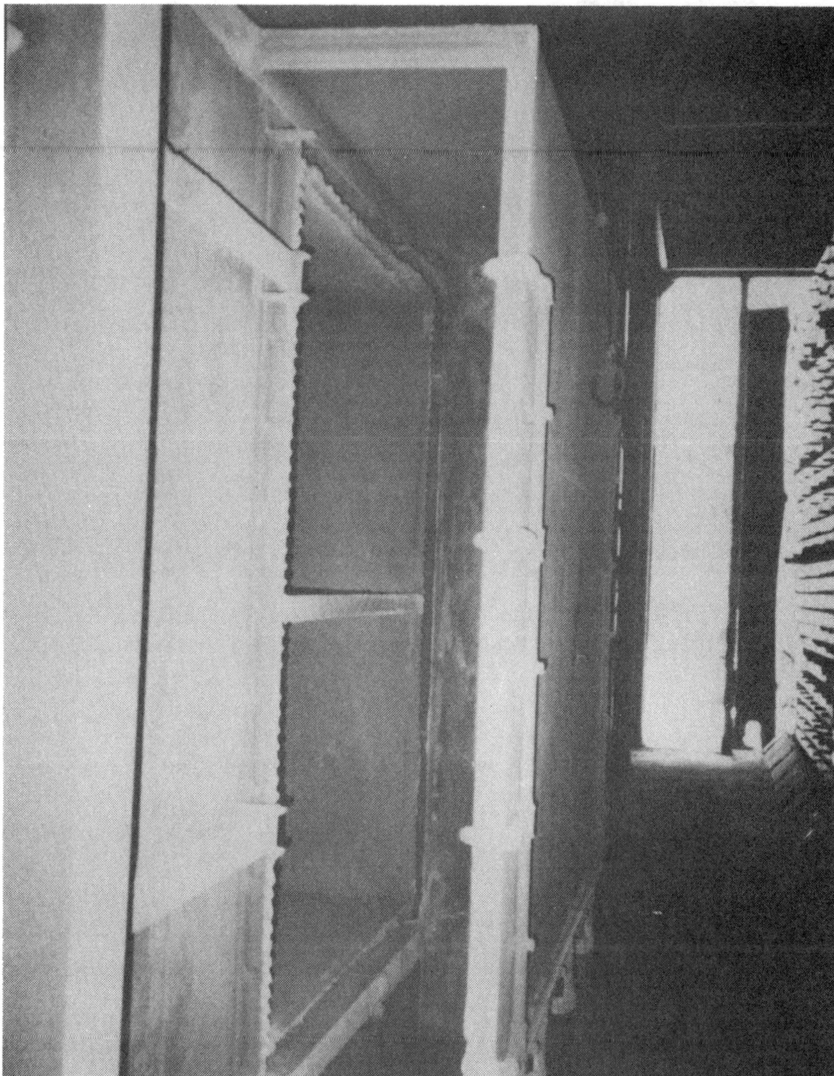

Figure 10.8. Single return air duct with radiation shield.

that fall within one of ten different velocity ranges and plotting them in their relative positions as viewed from the side of the unit. These symbols appear as variations in shading from light (high velocity) to dark (low velocity) in the printout. If your mind's eye could assign colors or shading to the air based on its velocity as it passed through the lumber, you might see something similar to these patterns.

Return air ducts need a shield in front of them to reduce heat radia-

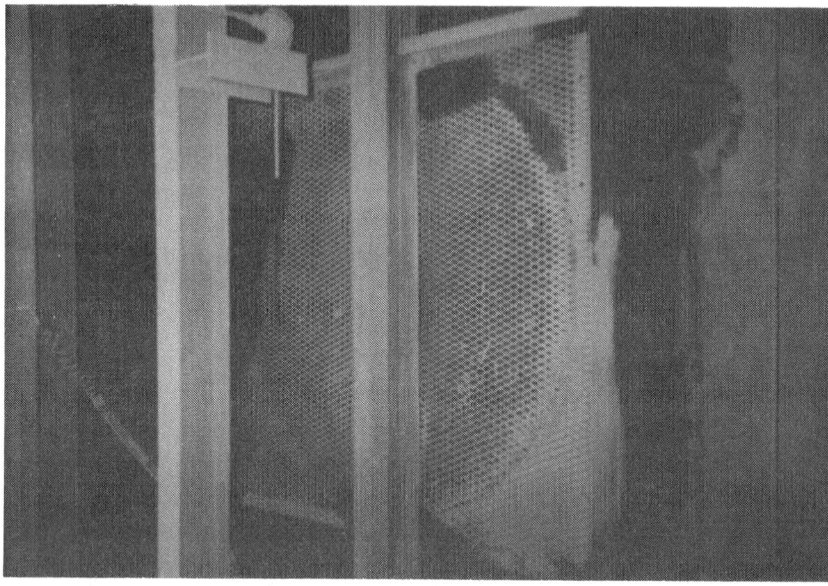

Figure 10.9. Single return air duct with wire mesh over the opening and no shield. Mesh this fine clogs quickly with ash and affects recirculation fan performance.

Figure 10.10. Single return duct with mesh replaced with 8 × 8 inch weld wire.

tion onto the lumber. These shields stand straight up or are angled against the wall. There does not seem to be any particular advantage to either design. However, air must be able to pass freely around its entire perimeter. A few manufacturers cover this opening with expanded steel mesh to prevent hard hats and kiln operators from being sucked into the blend chamber. Expanded steel mesh clogs rapidly with ash, reducing the efficiency of the recirculating blower. Replace this with 6 × 6 inch weldwire.

The solution to the uneven air migration problem is to split the single return duct into several ducts, spaced at regular intervals down the length of the kiln. This spreads out the withdrawal of air and minimizes the impact of the single large duct. A few mills experimented with splitting this duct by running the connecting ductwork within the kiln plenum chamber. These designs did not work well because they reduced the plenum chamber volume at the bottom, the worst possible place. The plenum chamber must be as wide as possible top to bottom. Otherwise, airflow is reduced in the bottom lumber layers, increasing our problems.

In 1983 I designed a split return duct that was installed on the outside of a kiln in Meldrim, Georgia. We mounted this duct between the kiln and the control room with openings cut into the kiln sidewalls. The improve-

Figure 10.11. Top view of a split return duct with four openings. Photo courtesy of McConnell Industries.

Figure 10.12. Radiation shield on a split return duct opening.

ment in airflow uniformity was dramatic. Since it was no longer necessary to overdry portions of the kiln to get other areas dry, drying time reduced and grade improved. This design has since been adopted by a kiln manufacturer for his direct-fired kilns. Figures 10.13 and 10.14 show the impact adding split return ducts had on the air velocity in both fan directions for

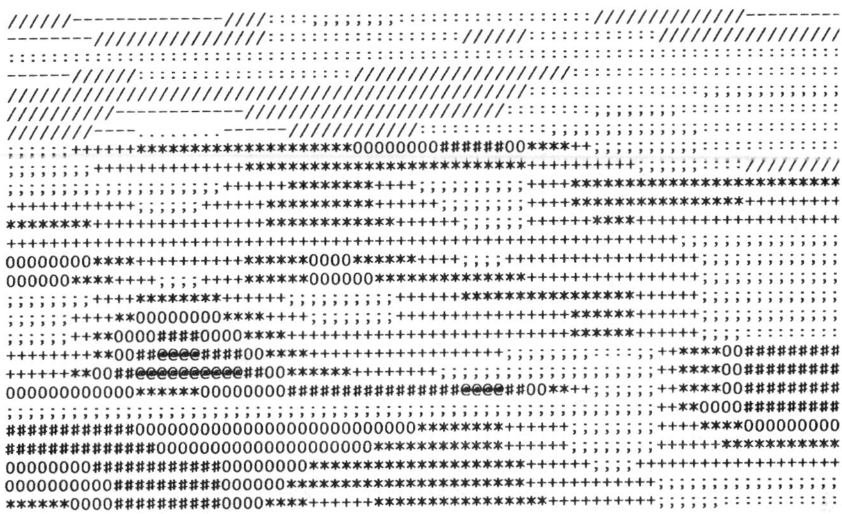

Figure 10.13. Air velocity profile in the same direct-fired kiln in Figures 10.6 and 10.7, but now fitted with a split return duct. Air blowing toward split return duct side. Average velocity is 1,110 fpm with a velocity range of 490 fpm.

Figure 10.14. Air velocity profile in the same direct-fired kiln in Figures 10.6 and 10.7, but now fitted with a split return duct. Air blowing away from split return duct side. Average velocity is 1,028 fpm with a velocity range of 560 fpm.

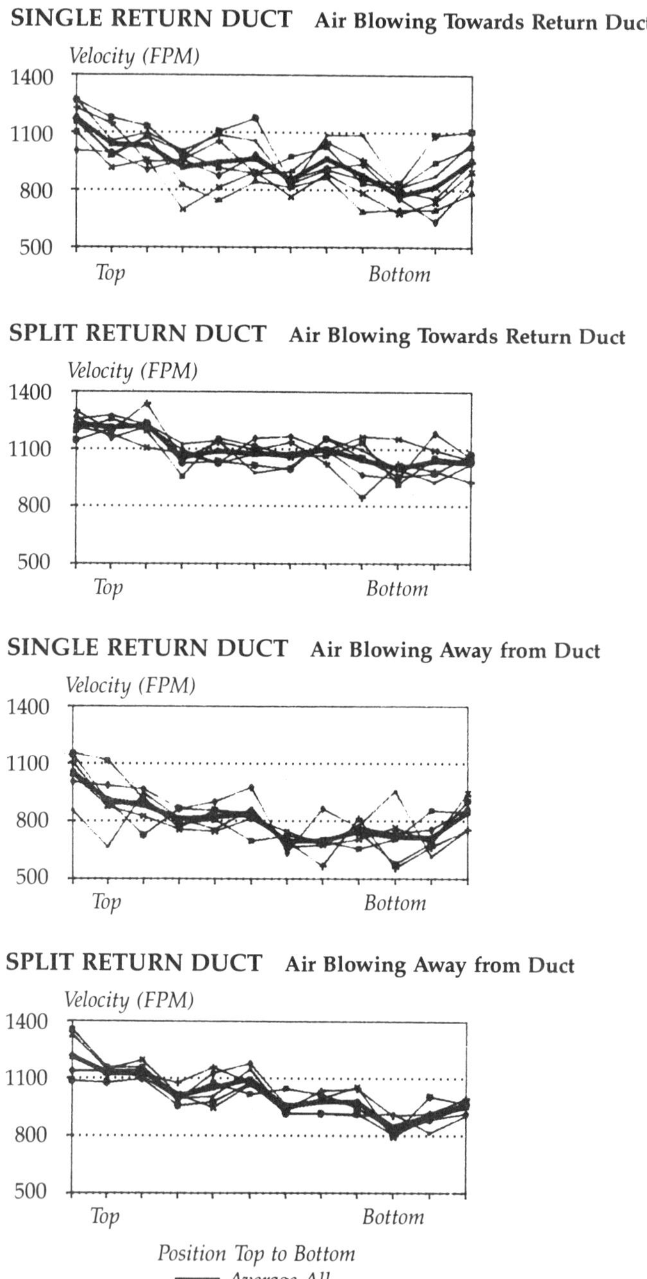

Figure 10.15. Comparison of air velocity data from Figures 10.6, 10.7, 10.13, and 10.14. Note the narrowing of variation that occurred when the split return duct was installed. Increase in average air velocity is not attributed to installation of return air duct.

the kiln profiled in Figures 10.6 and 10.7. Figure 10.15 graphically compares this data in both fan directions both before and after installation of the split return ducts. Notice how the variation has narrowed in both fan directions with the split return duct.

Still, the split return duct does not address the velocity drop that accompanies changes in fan direction. Most kilns see velocities drop around 100 fpm between fan directions. A few attempts have been made to install return ducts on both sides of the kilns. Dampers switch between the two ducts with fan direction. I have no information on the success of these efforts and am unaware of any in present operation. In theory, placing a split return duct on both sides of the kiln would give the most uniform drying for these kilns.

Handling Ash Deposits

Ash buildup is the most common complaint about direct-fired lumber. The smoky grey appearance it imparts to the lumber surface is objectionable to customers purchasing rough stock. This includes customers in the export markets and treating plants purchasing dry, rough timbers. Since most sawmills realize this, these kilns are usually built only at mills that plan to dress all of their production.

This ash causes problems at the planing mill since it shakes off the lumber as it is unstacked, conveyed, and fed to the planer. The health impact on planing mill workers is only now gaining attention. New OSHA standards regulating airborne dust requires its elimination from the workplace. Personal protection devices are not acceptable under the new rules. This places a tremendous burden on the employer to remove or control this dust before worker exposure.

Anyone who has visited a planing mill that dresses this lumber is immediately struck by the fine layer of dark dust on everything. Most managers recognize this problem and attempt to deal with it in a variety of ways.

The most common method involves spraying a fine water mist on the lumber as it is unstacked to prevent the dust from becoming airborne. This works fairly well, provided provisions are made to capture and remove the small amount of dust that shakes off the underside of the lumber. The working area around the unstacker tends to become very dirty. Water runoff must be collected, settled, and recycled to avoid the discharge restrictions placed on process water in industrial plants. Water absorption by the lumber does not tend to be a problem because the time it remains on the surface is very short and the surface is dressed off.

A few mills have placed hoods directly over the unstacker in an attempt to suck dust off the lumber and out of the planing mill. In general

Figure 10.16. Accumulation of ash on direct-fired lumber.

this only collects the dust that actually becomes airborne in the vicinity of the unstacker. General working conditions tend to be dusty all the way to the planer. A variation on this is to enclose the entire unstacker, infeed

Figure 10.17. One method to deal with this ash is to mist it with water at the unstacker.

conveyors, and planer. This significantly reduces the dust problem outside these areas and appears to be the most efficient method for handling the problem.

Experiments have been made using a rotary brush system ahead of the planer. This cleans up the shavings but does not address the airborne dust problem that occurs ahead of it. Actually this method has some merit when used on a continuous basis. Direct-fired lumber tends to become darker the longer a kiln is in operation, particularly in water spray systems. New kilns produce bright lumber with little ash. Since planer shavings are used as fuel, ash tends to recycle and build up in the system. Brushing ash from the lumber before it is planed reduces this buildup.

One mill is experimenting with a blower system mounted at the kiln doors. It is designed to blow the ash off as lumber is pushed from the kiln. This system has not been in operation long enough to judge its efficiency. There is also no information on how this airborne dust will be regulated, if at all.

Manufacturers of these burners experiment with different methods to reduce ash at the kilns. Their efforts include slotting the outside of the burner at the choke collar to sift out incompletely burned fuel particles, and extending the length of the blend chamber to increase its volume and allow ash to settle out. All have met with some, although far from complete, success. The search goes on.

Direct-fired Control Systems

Almost all direct-fired kilns are set up with single-zone control. A few have been built with two zones, in effect dividing the kiln lengthwise into two single-zone kilns. These kilns sometimes experience control problems in regulating and balancing heat distribution to the two ends.

With single-zones, most direct-fired kilns use traditional pneumatic controllers. These may be RTD or the older gas-filled systems. A few manufacturers use modern digital controllers with thermocouples as sensors. They may tie several thermocouples together and space them down the kiln length to get an average control temperature. This is not recommended by the thermocouple manufacturers.

Since most direct-fired kilns monitor supply and return air duct temperatures, this temperature difference can be related to lumber dryness. This works similar to the TDAL (temperature drop across the load) method for determining when to stop the kiln. Fan direction is important when using this method to pull the kiln. Always start your kiln cycle with the fans blowing in the same direction to get consistent, reliable data. You will get to setpoint faster by starting cycles with the fans blowing air through the lumber toward the return air duct. Give them a full 3 hours in that direction.

I have seen several ideas for zoning these kilns into more than one or two control zones. A few attempt to regulate heat top to bottom within the kiln similar to multizoned steam kilns. None have been implemented. Be very cautious in considering these proposals—there are some very difficult technical problems. Until now, designs have been limited by the service conditions actuators must endure in these kilns. I expect this limitation will be overcome by good kiln design or the availability of actuators designed for severe service. The real problem is more difficult. Since all control heat is delivered by a set air volume carried through interconnected ductwork, any change in air/heat delivery in one area results in additional air volume somewhere else. Close a damper here, and its air appears elsewhere with unpredictable results. Balancing control to the different areas of the kiln will be complicated, risky, and expensive. These systems will doubtless approach steam kilns in cost, without their reliable maintenance and quality track record.

Here are a few proven suggestions to help your system run more efficiently:

1. Clean the kiln on a regular basis, paying close attention to ash buildup on duct outlets and inlets. Extremely dusty systems require washing once a week. Clogged delivery systems extend your drying time and increase lumber moisture variability.
2. Scrape the ash buildup from your fans and recirculation blower at least once a month, more frequently if the buildup is heavy. Allowed to accumulate, chunks break loose, unbalancing the blades and causing premature failure. If you are losing a lot of fans, this may be the reason.
3. While you change out charges, leave the burner on. This saves gas, extends the life of burner refractory, and shortens kiln heatup time.
4. You will get to setpoint faster by starting cycles with the fans blowing air through the lumber toward the return air duct. Give them their full 3 hours in that direction.

CHAPTER 11

High-temperature Drying Controls

As I grow older, I pay less attention to what men say. I just watch what they do.

Andrew Carnegie

As a lumber inspector for SPIB during the latter part of the 1970s I visited many lumber mills throughout the South. I was constantly struck by the disregard many of them felt toward their drykilns. Kilns were a necessary evil, a black box into which they placed green lumber and removed dry. Poor grade recovery was accepted. Few mills appreciated the real impact good drying would have on their sawmill. After all, the real business was sawing lumber and production was king. Chip-n-Saws and double arbor gangs were exciting additions to the sawmilling trade. All problems could be solved with just a little extra production.

Rising log costs began to get everyone's attention as the new decade began. There was a sudden scramble to reduce saw kerfs and install scanners to reduce waste and improve log breakdown, board edging, and trimming. A few mills began to realize that the true value of their lumber lay in what made it out of the back door. Their grade recovery was heavily influenced by what happened to the lumber in the kiln. The kilns, in turn, were heavily influenced by how they were designed, operated, and controlled.

In 1982 an evolution in the control systems for high-temperature kilns began. That year saw the introduction of computerized, multizone drying controls on two kilns at Union Camp Corporation's Franklin, Virginia, pine sawmill. The benefit of these systems has since extended to all lumber species, since the need for good drying control is universal. These systems offer benefits in grade, energy, and drying time reductions far in excess of their initial investment. They are truly the controls of the future. Their use will only increase in the coming years.

Figure 11.1. Long shot of the first computerized, multizone drykiln. Photo courtesy of Union Camp Corporation.

Computerized Controls

The original computerized system used a PLC-type process controller, which programmed in ladder relay logic. This controller printed process data at regular intervals and would alarm motor and sensor failures. Interpreting the printed information required extensive knowledge of the process and equipment for clear understanding. Unprinted process data was lost at the beginning of each new charge.

This system was a resounding success for Union Camp. Overall grade recovery increased from an average 70.7 percent #2&Btr to 81.9 percent. Energy costs reduced 43 percent for steam and 10 percent for electrical power. Total savings including grade, energy, labor, and maintenance resulted in a 1.2 year payback on investment. In 1988 this computer system was replaced with advanced computer controls manufactured by a major kiln supplier, but still patterned after the original. The percentage of #2&Btr increased an additional 2 percent, with a slight reduction in drying time.

These systems use a combination of color monitor display, printer, plotter, and 20-MB hard disk designed for total system management and control. The monitor displays real-time process information updated several times per minute. These displays include a status screen showing the control temperature in each control zone as well as all-important drying information. Cross-sectional screens display the temperatures in each section of the kiln, all in their relative positions. A graph of the current

Figure 11.2. Comparison of green lumber and dry lumber from a computerized, multizone drykiln. Aside from color and stack height, there is little to distinguish the two. Top layers retain their flatness.

charge tracks average drybulb, wetbulb, moisture content, steam pressure, and fan speed. The home screen displays pertinent information on all kilns on the system. Maintenance screens allow the operator to test all control subsystems from the terminal.

Alarms are immediately announced on the monitor and printer and require physical acknowledgment by the operator. Important screen information may be printed out on demand and saved for future reference. Periodic status reports are printed automatically at key intervals during the charge. At charge end graphs and important data are saved automatically to hard disk for later screen reference or retrieval on the six-pen plotter.

This level of control and data handling sophistication means that complete drying information is presented in a clearly understandable form. This allows quick detection and isolation of equipment failures, reducing their impact. Schedules are easily optimized for time, quality, energy use, or all three at once. Seasonal changes requiring adjustments to schedules become apparent. Operator abilities improve as they accomplish their jobs faster and with greater accuracy and security.

Expanded computer power has enhanced the structure of kiln schedules. Not only does the computer store and retrieve far more schedules than before, the schedule format has the flexibility to handle any timber species. The ability to make easy schedule changes encourages experimentation among different drying strategies.

Drybulb and wetbulb setpoints ramp up or down between programmed setpoint intervals. Fan speed becomes another drying parameter manipulated to best advantage. Operators select among conventional, exhaust, and demand venting to maintain humidity control.

To enhance system reliability the computer is protected by a surge suppressor and uninterruptable power supply. The computer detects power outages and remains on line as it places each kiln on hold and stores all run-time data to hard disk. Once power is restored the kilns restart automatically, picking up where they were before power interruption. Kilns may be started automatically at preset times in the future to accommodate personnel time schedules or anticipated increases in steam availability.

The application of variable speed fans to kiln drying is not a new idea but one whose time has come. Above the fiber saturation point (FSP) air velocity is very important to rapid moisture removal. Below FBP air velocity has little impact. The technology for reducing air speed has been available for over five decades, yet few kilns were found with this feature until a few years ago. The problem was simple: Any reductions in fan speed must be manually invoked. This is a notoriously unreliable way to do anything, particularly when premature velocity reductions above FSP extend drying time. Reduce speeds too soon and drying times extend; reduce speeds too late and you do not receive the full benefits of variable-speed fan technology.

The computer has removed timing as a matter of concern. Speed changes are programmed to occur automatically, without operator intervention. They can respond to time or measured changes in moisture content, adapting speed to moisture level. Fan speed becomes another control variable like drybulb and wetbulb setpoints, adjusted at will. Relieved of the technical burden of using variable-speed fans, the operator is free to explore the limits of its use. The benefits are substantial, with electrical energy savings up to 45 percent.

Rising energy costs and growing concern over the environmental impact of combustion products have become a permanent part of business. We have already discussed the energy-saving potential reducing fan speeds bring. Computerized controls also allow us to manage our steam supply. Ramped startups and setpoint changes control boiler surging and reduce the belching black smoke commonly associated with these transitions. By monitoring boiler pressure the computer is able to sense prob-

Figure 11.3. Control room of a modern computerized, multizone drykiln. Photo courtesy of International Paper.

lems at the boiler that affect steam availability. Once detected, the computer begins cutting back on the steam demand of individual kilns on a priority basis. Kilns with the lowest priority gradually reduce their steam demand to allow boiler recovery. As steam pressure is restored, kilns gradually return to the system in reverse order of priority. Mills with marginal boiler capacity often use this feature to maintain a higher level of steam pressure than possible under normal conditions.

This is an integrated computerized kiln control system, linking the kiln control functions into the energy systems that affect efficient operation. Some manufacturers go so far as to add inventory control functions to their systems, taking this integration a step further. Do not lose sight of the fact that the primary function of a control system is efficient control of drying conditions within the kiln. A system that does an exceptional job arranging inventories may not give the best control. Be cautious about systems that promise feedback from the planing mill to automatically adjust drying schedules. I am uncomfortable about removing the human ability to detect problems from the control loop. Combining functions is fine, provided each portion is the best it can be.

MOISTURE DISTRIBUTION

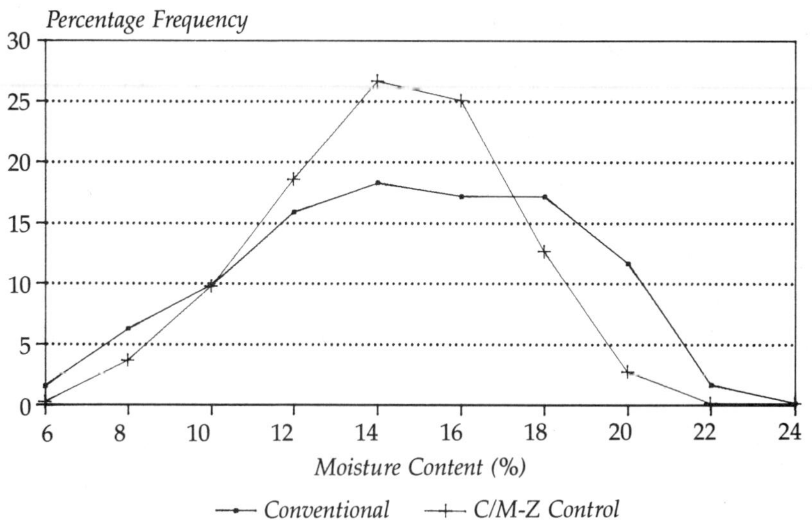

Figure 11.4. Change in moisture distribution at one mill installing a computerized, multizone drykiln. Information courtesy of Champion International.

After many years of operation covering hundreds of new kilns and conversions, we have data that serves as a guideline for determining potential benefits of computerized control.

Grade Improvement Conversion of existing kilns to computerized control results in a modest 2 percent to 5 percent improvement in grade. Conversion to computerized multizone control has seen improvements varying from 4 percent to 12 percent. Individual results depend on the type of drying done before conversion, whether the kiln was converted using the same drying methods, or if controls alone are replaced on an existing system.

Moisture Distributions Moisture distributions narrow in kilns converted to computerized, multizone controls. Simply converting to computer controls without multizoning is not known to have any impact on moisture distributions. See Figures 11.4 through 11.6.

Drying Time If you are converting existing kilns to computerized multizone controls, expect to see drying times reduce no more than 10 percent for high-temperature southern pine. Longer-drying species using

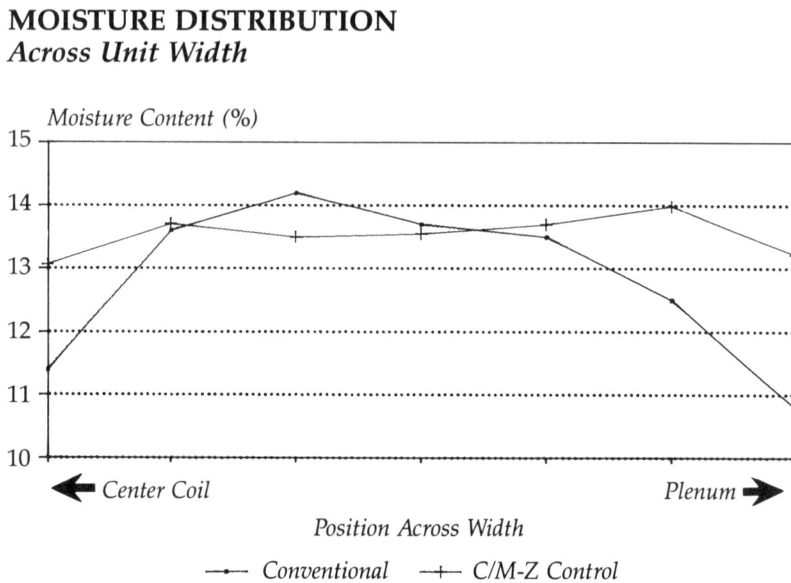

Figure 11.5. Change in moisture distribution across the width of a unit from installing a computerized, multizone drykiln. Comparison made between different kilns running concurrently. Information courtesy of Champion International.

conventional schedules have seen reductions to 40 percent. When controls alone were substituted, I have noted little change in drying time.

Energy Energy savings through conversion to computerized, multizone drykilns range from 17 percent to 43 percent.

Your conversion results depend entirely on the technology of your individual supplier. Talk to, and visit, customers of potential suppliers of this equipment before committing to a decision. If you have any doubts, require a reasonable performance guarantee.

The computer does not replace good drykiln practices or good drykiln personnel. The greater the understanding and ability of the people who use it, the greater the potential benefits. It does not think, but only reacts to situations as it has been told to react. It cannot compensate for things you forget. Do not expect the computer to do everything. There is no universal drying schedule suited to every kiln. Each is unique in some way. It is the responsibility of the operator to discover what works best for his situation.

Computerized controls are mandatory for multizoned kilns. A few

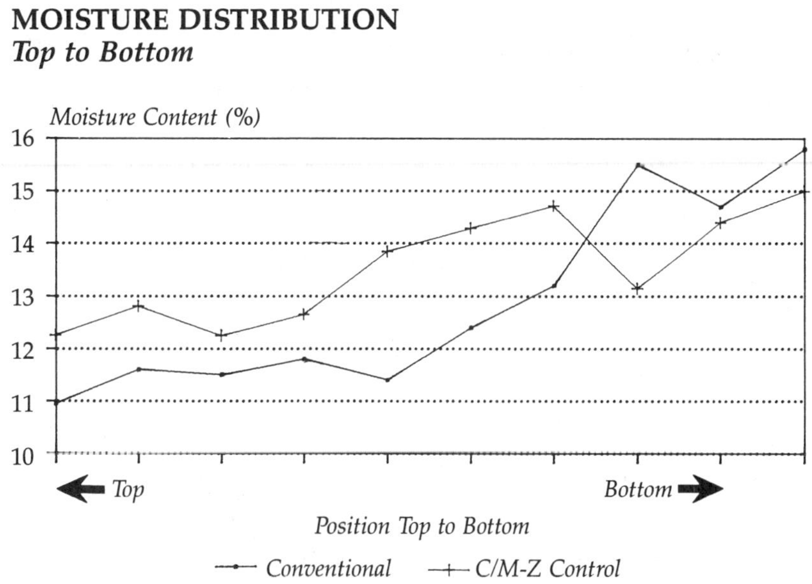

Figure 11.6. Change in moisture distribution top to bottom from installing a computerized multizone drykiln. Comparison made between different kilns running concurrently. Information courtesy of Champion International.

manufacturers have attempted to multizone and used several pneumatic controllers to run the kiln. These installations miss the point of multizoning. It is difficult enough to keep these controllers in calibration with themselves. Keeping them in calibration with each other is a challenge. Small variations in calibration or setpoint magnify when several control loops are involved. This increases the heat variability in the kiln, with the result that some areas of the kiln dry faster than others.

Adding computerized controls to an existing kiln without altering the kiln itself brings modest benefits at best. You will gain some advantage through better control and schedule options, provided you exercise them. Simple substitution of controls may not pay back your investment. Unless you modify the way your kiln performs in terms of heat delivery, air flow, or control method, your benefits will not approach the full capability of the equipment. Real changes bring real benefits.

These systems have only been used on steam kilns due to their capability for multizoning. Direct-fired kilns are also candidates for computers, but their benefits are not proven. Their potential for benefit is through increased schedule options and control. They also offer enhanced information-gathering capability for maintaining heat balance in these kilns.

MOISTURE DISTRIBUTION
Comparison of Drying Methods

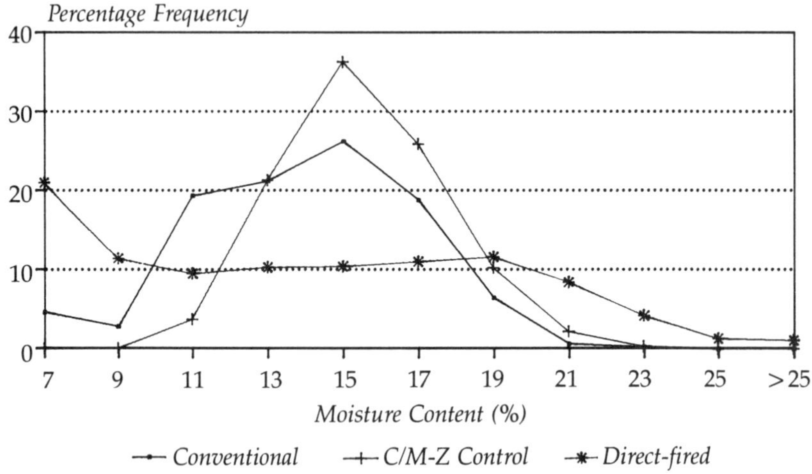

Figure 11.7. Comparison of moisture distributions between three drying systems. Data from direct-fired kiln is from a kiln with a single-return duct.

The following sections describe alternative controller options available to single- and double-zoned kilns.

Vapor-filled Pneumatic Recorder/Controllers

This is the workhorse controller for conventional kilns. I have a model that was patented in 1894, so they have a long record of reliable service. These were the first controllers used on high-temperature kilns.

These controllers record and control temperatures by reacting to the pressure a sealed liquid exerts as it responds to temperature changes. The sensing element, which is a small cylindrical bulb, connects by capillary tube to a coiled, helical tube (Bourdon tube) within the controller. The entire assembly is sealed and filled with a liquid, which vaporizes and exerts pressure in relation to its temperature. As the temperature rises, rising pressure tends to straighten the Bourdon tube. Through mechanical linkages within the controller, this movement raises the pen indicating temperature and closes the control valve. With the pressure drop accompanying falling temperatures, the tube relaxes, lowering the pen and

Figure 11.8. Vapor-filled pneumatic recorder/controller installed on a steam-heated kiln.

opening the control valve. This describes a simple single-bulb controller. The wetbulb and each control zone has its own control assembly consisting of bulb, capillary, Bourdon tube, with indicating and control linkages. However, the wetbulb control works in reverse, opening above setpoint and closing below.

High-temperature kilns require split drybulbs. These controllers are designed with two bulbs for each control zone. The capillary tube begins at the controller as a single tube, but soon splits into two capillaries running to separate bulbs. The junction of these capillaries has a pressure-sensitive valve that opens the tube with the highest pressure and closes the other. In this way the controller always sees the highest pressure and therefore reads the highest temperature of the two bulbs. With changes in fan direction, the controller automatically switches to the highest temperature for control. Without that pressure valve between the bulbs they read an average temperature for the two bulbs.

This brings us to the major drawback of these controllers. Because they always measure the highest temperature between the two bulbs, they only control on entering air temperature—the temperature off the coil before the heated air enters the lumber. Better grade recovery is sometimes achieved by using exiting air control, which controls temperature as it leaves the lumber, after it has done its work. It automatically adjusts incoming temperatures to the dryness of the lumber the heated air passes through.

These controllers operate as proportional controllers within a percentage band about setpoint. The farther below setpoint the temperature falls, the more the control valve opens. The farther above setpoint the temperature rises, the more the control valve closes. In operation the controller constantly overshoots and undershoots setpoint after you reach control temperatures, oscillating within 5° F. The constant oscillation about setpoint is inefficient and adds to drying time. During warmup prior to reaching setpoint, these controllers deliver the maximum available heat. As it begins controlling around setpoint, temperatures begin to swing along with boiler demand. Toward the end of the schedule temperatures may override setpoint. This is a limit in the calibration and tuning capabilities of these controllers.

Loss of temperature sensors in these controllers requires major repair. Moore units in the familiar green case are returned to the factory for bulb replacement. This is not a field repair—you must keep a spare controller on hand. Capillaries must be carefully removed from the kiln and packed with the control unit for shipping. Be very careful when handling and installing these units. Kinks in the tubing render the bulb useless. Foxboro controllers are repairable in the field by knowledgeable service representatives.

Calibration of these units is traditionally done using a hot-water or hot-oil bath. Calibrate high-temperature kilns with hot oil, which will allow you to calibrate at your control temperature. Hot oil carries some risk, so handle carefully. Immerse the bulb completely in the hot oil and compare the temperature of the oil to the temperature shown on the controller. They should agree. If not, calibrate according to manufacturer's instructions.

A spot check of calibration may be done with a stainless steel thermos of boiling water. I use hot condensate. Immerse the bulb and a calibration thermometer in the thermos and keep it agitated. Take a temperature reading and compare it to the controller. Record it. As the temperature drops, continue to compare the thermometer to the controller, recording the drop for 30° F or more. Plot the thermometer readings against the controller to view your calibration curve. The unit is calibrated if the temperature of the bath always corresponds to the recorder readings.

Figure 11.9. Vapor-filled pneumatic recorder/controller installed on a direct-fired kiln.

It is possible to calibrate your controller while the kiln is running. Obtain an electronic temperature meter designed to read thermocouples. Run thermocouple wire from the control room into the kiln to the bulb you wish to calibrate. At the bulb strip the wire leads back, twist them together, and crimp in a sleeve. Wrap this around the bulb. In the control room simply attach the other end to the meter. This method permits calibration at operating temperatures and is much safer than hot oil.

Calibration Procedure With the kiln running and at setpoint, compare the thermocouple temperature to the recorder temperature. The controller pen must be tracking at setpoint. Adjust the controller pen arm to the average meter temperature using the fine screw adjustment on the arm. This calibrates temperature. Adjust the setpoint indicator to the new track temperature by using the fine screw adjustment on the indicator arm (Foxboro) or by repositioning the needle on the setpoint dial (Moore). This calibrates setpoint. Move the setpoint to the desired level using the setpoint adjustment knob. Once the kiln settles at the new setpoint the controller should track with the thermocouple meter.

These controllers are rugged and take a great deal of abuse, although they will last longer if mounted in a dry environment away from vibration.

Electronic/Pneumatic Controllers

Electronic/pneumatic controllers are similar in design to the traditional vapor bulb controllers except that they use RTDs to measure temperature. The controllers are very reliable and have the advantage of permitting you to select entering or exiting air control for your kiln. Switching sensors during fan reversals is done by a relay tied into the fan starter system.

These controllers offer more stable setpoint control than the older style controllers that only have proportional control. Most have adjustments for proportional band and integral gain to reduce valve oscillation and track setpoints. Since dry kilns respond very slowly to changes in heat delivery, I end up setting these very low, particularly when the controller is set up for exiting air heat control.

Maintenance is much simpler. Bulb losses are repaired by simply replacing the bulb. These bulbs are connected to the controller by three lead wires run in a copper shield. Some manufacturers sell you new replacement bulbs with new full-length shielded lead wire. At $100 or more for the RTD and $3.50/foot for new leads, the money mounts up fast. There is a way to save money on replacement parts. Purchase off-the-shelf 100 Ohm RTDs from a reputable supplier such as Omega Engineering. These come in different sheath lengths with and without connector heads. Get a catalog and select the assembly that suits your needs. The three lead wires are not special or unique to any individual RTD so they can be reused. The next time an RTD goes bad, cut it off at the sensor end and run the shielded wire directly into the new RTD connector head. Some people buy RTDs without connector heads and use ½-inch LB electrical boxes for wiring connections. Use a rubber compression fitting to secure the shield to the head entry. Make the wire connections in the

Figure 11.10. Electronic/pneumatic controllers installed on a multizoned drykiln.

head. Keep in mind that the three leads are separate and important, even though two are the same color: If you bundle the two white wires together on one lead, the bulb will not work.

You can spot check calibration with the hot water method, although it is unusual for these controllers to get out of calibration. Purchase calibration resistors from the manufacturer to do your own calibration. One resistor calibrates at 100° F, the other at 250° F. I rarely find these units out of calibration. Something else is usually wrong.

One mill told me they could not keep their controllers in calibration. Every time the service representative for the kiln manufacturer came in to calibrate the controllers, they simply could not dry lumber. They followed his setpoint instructions but always returned to running the kilns with the valves wide open and out of control. He had just left and they were no better off than they were before. I visited the mill and found that the controllers were in calibration. The service rep had used the calibration resistors, given his blessing, and gone on his way. He had neglected, however, to determine if the controllers were connected to anything. Falling boards had dislodged and destroyed the temperature sensors. Investigate all possible causes of failure.

The relay that switches between sensors during fan reversals often fails at the worst possible time. Keep one on hand at all times.

Mount the controller on a vibration-free surface away from the heat of the kiln. The control room atmosphere must not allow moisture to condense on the control unit.

Electronic Digital Controllers

A few manufacturers have begun supplying several single-loop digital controllers to run their kilns. These units use either RTDs or thermocouples for temperature sensors. Beware of attempts to tie several thermocouples together in parallel to read an average temperature. Although this works in theory, the actual average may be different than what you expect. This is not recommended by the thermocouple manufacturers. None of these units switch internally between sensors for fan reversals. This must be done outside the unit by the supplier. Incorrectly done, this is a potential source of control signal error.

Figure 11.11. Electronic digital controllers installed on a direct-fired kiln.

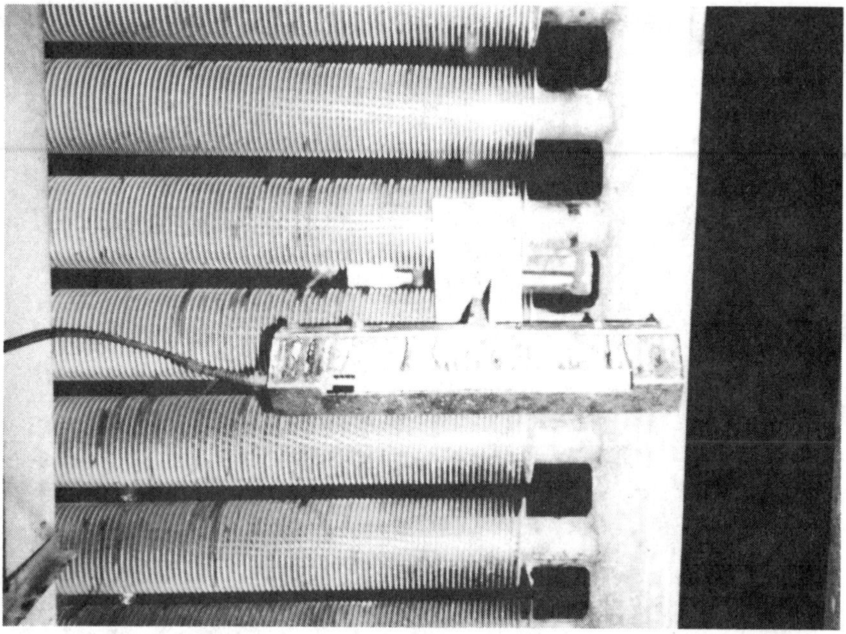

Figure 11.12. Sensor mounting method on the center coil for vapor bulbs and electronic sensors. Bulbs must be shielded from heat radiation off the coil.

Beyond this, these are excellent controllers capable of precision temperature control. Some are programmable for different schedules, but I would not use them as a substitute for computers in multizoned kilns. There is too much room for error when programming multiple controllers. One drawback is that you will need a separate recorder if you want to chart temperatures. This usually requires a separate sensor network.

Mounting Drybulbs

There are two ways to install drybulbs in your single-zone kiln. Most single-zoned steam kilns mount the split bulbs under the overhead coils, one in each plenum. We know from multizone kilns that the overhead coils are often the first to reach setpoint. This means that we may restrict the heat we need to the center coils and extend drying time. Placing the split bulbs on either side of the center coils about 4 feet off the floor will deliver more heat to the center coils and speed drying. Watch your grade recovery, though. Direct-fired kilns mount their bulbs in the outside plenums. Place them midway of the kiln length.

Two-zone kilns divided into zones along the kiln length have the same options for installing their bulbs as single-zone kilns. Place the

Figure 11.13. Stationary drybulb mount in the outside plenum. There is some danger of damage from stray boards when you are pushing and pulling the kiln.

bulbs midway of each zone. Kilns divided into overhead and center zones combine the two mounting positions for single-zone kilns. They mount the bulbs for the overhead coils under the coils in the outside plenum. The center coil bulbs mount on both sides of the center coil.

All center coil bulbs must be shielded from coil radiation and protected from falling boards. I prefer heavy, rigid steel supports to deflect boards.

Never place bulbs above the overhead coils or between double center coil banks. This only reads the heat contribution of one coil. Bulbs are protected from physical damage, but your schedules will be entirely unique to your operation. It will be difficult to compare notes with other operators. This method gives poor control and increases degrade.

For multizone systems controlling on exiting air, it is important that sensors mount close to the lumber, 6 inches to 9 inches is a good distance. The first systems mounted these bulbs on rigid supports, but this often presented problems when pulling the kiln. Mounting them on swing out arms eliminates this concern and also permits you to adjust placement to fit load conditions.

REFERENCES

1. Culpepper, L. G. "Application of Computerized Controls to Improve the Lumber Drying Operation." Seattle, WA: IUFRO Meeting, 1989.
2. Foxboro. *Installation, Operation, Maintenance Instructions for Multizone Dry Kiln Controller with Duplex Wet Bulb Control*. Foxboro, MA: Foxboro, December 1977.
3. Irvington-Moore. *Instruction Manual for Round-Chart Recording Controllers, Models 4333-10W, 20W & 30W.* Jacksonville, FL: June 1980.
4. Sanders, C. M. *Inspection and Maintenance of High Temperature Dry Kilns and Equipment*. Jacksonville, FL: Irvington-Moore, October 1986.
5. Wengert, E. M. "Trouble-shooting a Pneumatic Recorder/Controller." *FPRS News Digest*, File F-3.20 (April 1978).

CHAPTER 12

The Drying Process

A great pleasure in life is doing what people say you cannot do.
Walter Gagehot

A well-run high-temperature kiln, using the latest technology, far surpasses other large volume drying processes in quality and economy. Poorly operated, however, it ruins a lot of lumber in a short period of time. Understanding how the lumber dries in these kilns is fundamental to assuring their proper operation.

About Southern Pine

Southern pine wood is a porous material composed of countless microscopic cells called longitudinal tracheids. Tracheids are best described as closed-ended hollow tubes whose lengths are a hundred times their diameter. Countless millions are arranged lengthwise in a tree stem and cemented together with lignin to form wood. This direction along the stem length is known as the *longitudinal* direction in wood. By chemically removing lignin to release these fibers, we can form the fibers into thin mats to produce paper. Drying wood fibers in these mats is infinitely easier than drying them as a solid block of wood.

Not all cells are arranged with the length of the tree. A small percentage radiate from the center of the log, the *radial* direction in wood. These are called wood rays. Have you ever noticed small checks in the ends of your lumber when you open the kiln a little too soon? These small end checks follow the direction of the rays.

A third direction in wood is called the *tangential*. Imagine the end of a log. Draw a line straight up from its center in a radial direction. Now, make a "T" by drawing a line across the top. This line is the tangential direction in wood. The three directions–longitudinal, radial, and tangential–will become more important to our discussion later. The orientation of these fibers as they appear on the lumber surface is described as its grain.

As a tree grows tracheids are laid down one layer at a time by the cambium layer. This layer separates the wood from the bark, producing wood to the inside and bark to the outside. As growth begins in the spring this layer becomes very watery and "slips" easily when peeled by debarkers. The spring, with ample water and stored nutrients, is a time of fast growth. Tracheids produced in the spring are thin walled and distinguished by their light appearance. This wood is called springwood or earlywood. As the year progresses into summer, water becomes less abundant and growth slows. The wood produced now, called summerwood or latewood, has thicker cell walls and appears much darker. These two layers of growth form one annual ring. We estimate the age of a tree by counting these rings on the exposed end of the log.

The width of these rings has an impact on the density of the wood. If growing conditions are favorable, the tree grows very fast and lays down very wide rings. It is not unusual to see rings ½ inch to ¾ inch wide or more. This lumber has lower density and lower strength. If a tree is forced to compete for space and nutrients, the rings are narrower and have a higher percentage of dense summerwood. I have a few samples of wood collected over the years that are very dense. One piece has over 80 rings per inch. Many operators comment to me that the wide-ringed material is very difficult to dry, that wet boards are often low density. This is not universally true. Indeed, high-density pieces are just as likely to have high moisture content. Although low-density pieces start out with high moisture content, they move water faster. Low moisture content high-density pieces, on the other hand, move this moisture more slowly. When checking moisture on the ends before pulling a kiln, ignore density.

The differences between heartwood and sapwood further complicate moisture movement. If you look at the end of a log you will notice that it has a dark central portion surrounded by a light band. The dark central area is heartwood. As a tree grows, fewer nutrients reach the central part of the tree and the wood cells die. In pine trees these cells often fill with pitch and other extractives, which darken the wood. This area is lower in moisture content than the lighter surrounding sapwood, which contains a high percentage of living cells.

Water in Wood

Freshly sawn lumber contains a lot of water. For example, the lumber in a fully loaded, 84-foot double-track kiln must evaporate over 30,000 gallons within a 24-hour period. This would fill a cube almost 16 feet on a side. Getting this water out without degrading the lumber is the ultimate goal of kiln drying.

Where is this water? The majority of the water is held in the cell cavities as free water. Free water is found in lumber with moisture contents above around 30 percent. This is known as the fiber saturation point (FSP). The moisture content at FSP varies between and within different timber species. Below FSP water is held in the cell walls as bound water.

Free water moves rapidly to the lumber surface, where it evaporates. Open an access door after the kiln has been on for about an hour and you will see water streaming off the lumber and pooling on the floor. A great deal drains out around the doors, but the majority evaporates. This water is pushed out of lumber by the thermal expansion of water and air within the cell cavities. It always seemed prudent to drain this water from the kiln to speed drying and reduce energy consumption. However, EPA standards for water treatment of process water suddenly become a factor when you do this. It is less expensive to evaporate this water than treat it. Reduce the amount of water that pools up on the floor by reducing humidity during rise to setpoint. After the thermal expansion stage free water continues to move to the surface by capillary action. In high-temperature kilns water literally boils out. High air velocity is required to remove this moisture from the surface. Most high-temperature kilns pass the fiber saturation point two-thirds of the way through the cycle.

Below FSP water is bound within the cell walls and is more difficult to remove. Its movement is by the slow diffusion of water from high (lumber core) to low (lumber surface) moisture content areas. This moisture movement is largely unaffected by fan speed and is the reason we can reduce air speed during the later part of the drying schedule. Wood begins shrinking as water is removed below FSP. If the wood is uniform in structure and all surfaces are exposed to the same drying conditions, this shrinkage has no impact. Any stresses produced by shrinkage on one side are offset by an equal shrinkage on the other.

Alas, if wood were but half so perfect! Wood has different properties in each of its three directions. All of these properties are related to the orientation of the cells. For example, moisture moves very easily within the cell cavities along the length of the tracheids—about nine times faster than through the cell wall. Also, the more cells oriented in any one direction, the less it shrinks in that direction. As a crude example, take a 1-inch section of plastic soda straw and try to crush it lengthwise between your thumb and forefinger. Compare that effort to what it takes to crush it across its diameter. Movement is easier across its diameter. No cells align tangentially and moisture movement is slow, while shrinkage is the greatest of the three directions. Radially, moisture moves a little faster because of the wood rays and shrinkage is somewhat less. Moisture movement is greatest and shrinkage the least in the longitudinal direction. On any board, all three fiber orientations may appear on any or all surfaces.

The local shrinkage differences as this lumber dries produces warp if unrestrained or otherwise controlled.

To understand how differences in shrinkage in the same board cause warp, lay a wide, green board on the pavement in the sun. As the exposed surface dries, it begins to shrink. The opposite side still remains near its beginning moisture content and does not shrink. As the differences in shrinkage increase, the board begins to cup on the dry side. As a cabinetmaker I use this technique to flatten dry, cupped shelving boards. The cupped board is simply laid on the grass on a sunny day, cupped side down. The sun dries the exposed surface while the cupped side picks up moisture from the ground. It does not take long for these boards to flatten out.

Still, we could probably deal with these shrinkage differences if they were all we had to contend with. Have you ever noticed the end of a log that had rings that were much wider on one side? These rings seem to bulge out from the center and are much darker in color than the rest of the wood. This is called reaction or compression wood, and it's a common growth-related defect found throughout the southern pine region.

Compression wood forms on the underside of stems that have bent over through some injury or outside force, such as a storm. It is the tree's response to the injury as it attempts to straighten itself out. We would not be concerned with compression wood except that it is very different in physical properties from normal wood. It shrinks much more in the longitudinal direction as it dries. Most kiln operators first become acquainted with compression wood when they open their kiln and discover boards bowed out wildly from the sides of units. They often curl and twist into bizarre shapes on the top. The next time you see this, examine the board carefully for the wide, dark-shaded rings of compression wood.

There has been a great deal of interest and controversy over the past few years concerning juvenile wood. This wood surrounds the center of a log and has different shrinkage properties from wood that forms later. This is feared to be a drawback to plantation-grown trees. I have never encountered a drying problem I could specifically relate to juvenile wood, although there is no doubt that I have seen and dried plantation material. It is possible that the problems blamed on juvenile wood in the southern United States stem from the schedules or control methods used to dry it.

Other, minor growth characteristics also cause variations in shrinkage. We are all familiar with the grain appearance of wood as it passes around a knot. These local grain distortions are a source of warp, usually removed by trimming. Spiral grain appears as fibers oriented diagonally across the board surface. When I see this lumber on the grading chain, I am reminded of a large pine tree that once stood near my home. The tree stood alone in a field and had been struck and killed by lightning years before. As the bark sloughed off and the wood dried, surface checks developed that twisted around the tree from bottom to top—spiral grain. I always marveled at the appearance of this tree.

The High-temperature Drying Cycle

With the kiln loaded we begin the drying cycle with the startup sequence. This sequence may be manual, semiautomatic, or automatic depending on control system. First, the vents open and the fans start, with the vents closing after the last fan comes on. This protects the kiln from implosion as hot kiln air suddenly cools and contracts in passing through the lumber. Drybulb and wetbulb setpoints are usually set high to allow the kiln to get to setpoint in the shortest possible time. This quick heatup does not harm the wood. We do, however, want to be at a setpoint of 235° F to 240° F measured on the entering side of the lumber within 3 to 4 hours.

Most kilns keep the vents completely closed during heatup and for several hours thereafter to allow the lumber to heat to its core before active moisture removal. This timing period varies from kiln to kiln and item to item. A good starting point is one-third of the total cycle length. It is actually advantageous to vent a small amount during this time to speed drying. I do this by setting my initial wetbulb setpoint 2° F to 5° F below what the kiln would naturally achieve with the vents closed. This is a heavy steaming period for the kiln, with steam escaping through every crack and seam. During this time the lumber becomes flexible, attaining the plastic quality that allows it to move freely. Misplaced sticks, cross-outs, and truck supports now begin to distort lumber at contact points. This degrade becomes permanent as the lumber dries.

About two-thirds of the way through the cycle the majority of the lumber begins to pass the fiber saturation point. If we had trouble maintaining drybulb setpoints before, they now begin to come into control. Steam consumption slowly reduces as the lumber dries. It is during this time that drying conditions become critical. Too high a wetbulb temperature produces dark lumber and may increase warp. Too high a drybulb temperature overdries lumber and reduces its strength.

The high-temperature drying process literally boils water out of the lumber. This requires that we maintain temperatures above 212° F. Temperatures below this affect drying quality in high-temperature kilns.

Entering Air Heat Control

Until 1982 all kilns were controlled using entering air heat control. The majority still do. These kilns measure and control setpoint temperatures of the air as it leaves the heating coils, but before it enters the lumber. Control setpoints vary from 235° F to 240° F depending on air speed and its impact on temperature drop through the lumber. This temperature drop is known as temperature drop across the load (TDAL). The TDAL varies from 30° F to 50° F up to fiber saturation point and then begins a slow, gradual drop to the end of the charge. This value is often used to determine when to pull the kiln. Figure 12.1 illustrates how temperatures behave as air circulates in an entering air-controlled kiln. These tempera-

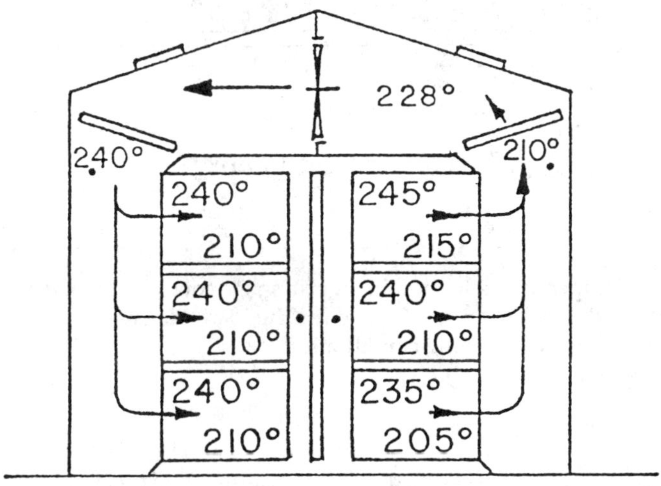

Entering Air Heat Control

Figure 12.1. Temperature variation in an entering air-controlled kiln controlling at 240° F with a 30° F drop across the load. Note the exiting air temperatures below boiling point.

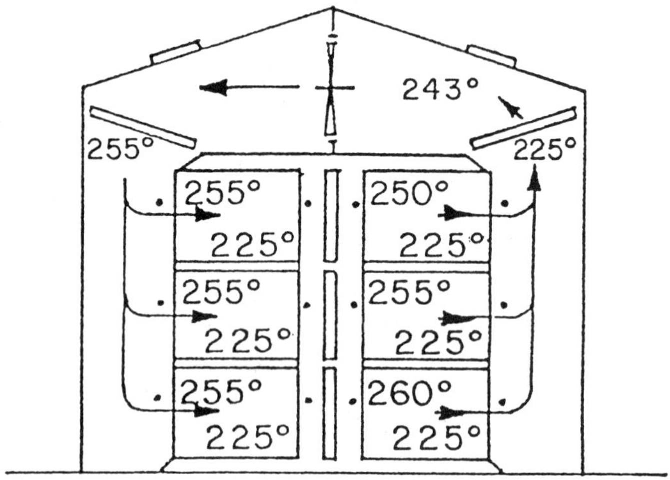

Exiting Air Heat Control

Figure 12.2. Temperature variation in an exiting air-controlled kiln controlling at 225° F with a 30° F temperature drop across the load. No area of the kiln is below high-temperature drying conditions.

tures are based on 240° F drybulb setpoint with a 30° F TDAL. As you see, temperatures fall below boiling point in several areas. Somewhere across most of the units of lumber we cease to high-temperature dry.

How temperatures vary throughout a drying cycle is illustrated in Figure 12.3, which compares the entering and exiting air temperatures of an entering air- and exiting air-controlled cycle. In 3 to 4 hours after startup we achieve setpoint of 240° F. Since the heat demand of green lumber is very high on startup, the TDAL will peak around 50° F before settling down in the 30° F to 35° F range. This means that the temperature of the air leaving the lumber is 205° F to 210° F—below high-temperature drying conditions. This condition persists through one-half to two-thirds of the

ENTERING AIR CONTROL

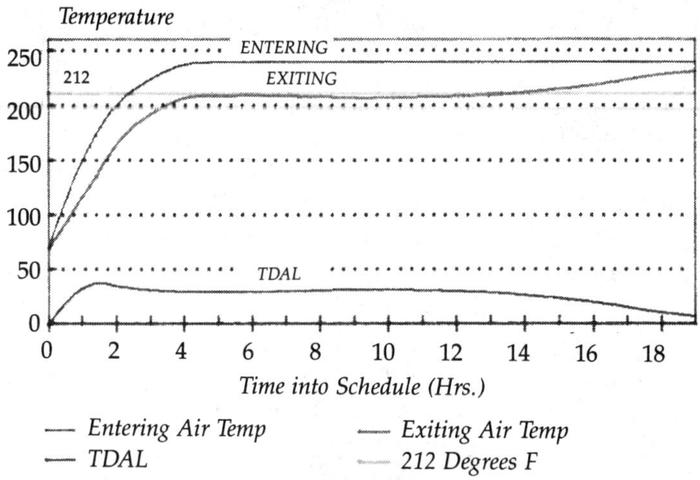

EXITING AIR CONTROL

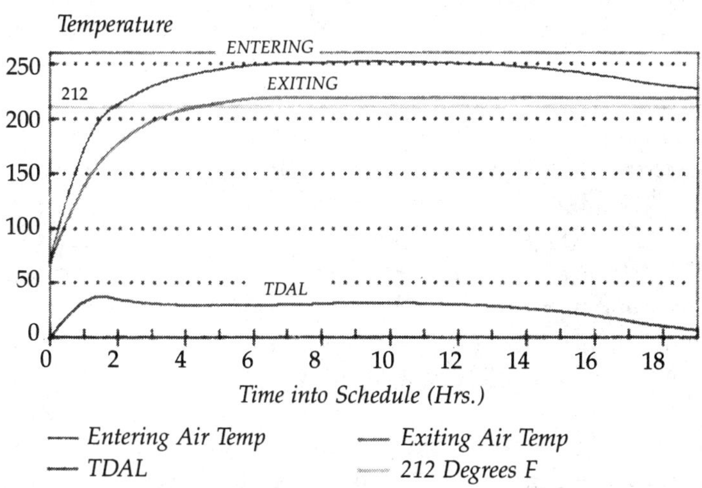

Figure 12.3. Comparison of entering air to exiting air control throughout the drying cycle. TDAL and cycle times are identical for comparison purposes. The boiling point is added for reference. The exiting air-controlled kiln is at high-temperature drying temperatures earlier and longer through the cycle.

schedule or until we pass the free water state and heat demand lessens.

If we are shutting down our kiln on a typical 8° F TDAL, we find the entering air temperature at 240° F with the air exiting at 232° F.

Exiting Air Heat Control

Exiting air heat control was designed into the very first computerized multizone drykiln. Although this was the first application for drykilns, this method is widely used in related industries for drier control. This method senses and controls temperatures after the heated air has cooled in passing through the lumber, as it exits the load. This method automatically adjusts entering air temperatures to the heat demand of the lumber it passes through. High moisture content lumber receives more heat and dries faster than lumber with lower moisture content. Operating setpoints vary from 215° F to 230° F with the average around 220° F. Figure 12.2 illustrates how temperatures vary in a multizoned kiln at an exiting air drybulb setpoint of 225° F with a 30° F TDAL. Here, entering air temperatures adjust as needed to maintain exiting air-control setpoints. Faster drying areas receive less heat, slower areas more.

Figure 12.3 shows a typical charge run at a 220° F exiting air setpoint and shows the impact on entering air temperatures. Although it takes 6 or more hours to reach setpoint in all zones, we are well above boiling point early in the schedule. With the temperature drop, entering air temperatures are in the 240° F to 250° F range. Some kilns approach 260° F or more. These temperatures early on in the schedule do not seem to cause degrade. The lumber is green and resilient to this heat.

These high temperatures continue until we pass the free water state and heat demand reduces at 12 to 14 hours. At shutdown our ending temperatures are 228° F on the entering air side and 220° F on the exiting. Much milder than ending conditions with entering air heat control—milder at that critical time when lumber is most prone to warp and degrade. Exiting air heat control delivers more heat during the early stages of drying to speed drying but stretches out the cycle toward the end. This widens your window of opportunity to pull the kiln when the lumber is optimally dry.

There is an added bonus to exiting air heat control that is hard to quantify. When starting up a new exiting air-control system I always watch the kiln operator to catch his reaction as he checks the first load with his resistance meter. As he drives in the needles he will often comment that the lumber couldn't be dry, the needles slide in too easily. Then he looks at the reading and is amazed to discover moisture contents far below what he expects. The lumber from exiting air-controlled kilns is much softer, and this may be an advantage for your customers.

One requirement of exiting air control is not widely recognized. Since we measure and control temperatures as the air leaves the load, it is important to have uniform baffling and loading from one charge to the next. Circumvention of air and heated air that is not representative of the true heat demand of the lumber cause erroneous readings. If temperature sensors align with a large gap between adjacent units the sensor will control on air that is artificially hot and reduce entering air temperatures accordingly. If a floor baffle falls below a sensor the reduced airflow through the unit demands more heat. If uncooled air from below the unit blows directly on the sensor it will demand less. Finally, mills that sort two or more lengths into the same unit will have problems if the staggered ends align with sensors. These ends have less lumber volume and also demand less heat. Expect the rest of the unit to be wet at the end of a normal drying cycle.

Determining When to Pull the Kiln

Kiln operators spend a great deal of gut-wrenching time trying to answer the all-important question of when to pull the kiln. They use a variety of methods. Alone, none are 100 percent reliable. The best results are achieved by using a combination of indicators.

When I first started into kiln drying I met a gentleman who used his pocket knife to tell if the lumber was right. First, he would tap the lumber with it, listening for the magic hollow sound of dry lumber. Next, he began cutting a sliver off the corner of a board and continued pulling it off with his fingers. How easily it pulled and how far were important indicators. Finally, he broke the sliver, listening for the snap and examining the broken ends. Fifty years of sawmilling experience told him if the lumber was "right" or not. I never learned this trick and regret not paying it closer attention. The only problem with it was you had to have physical contact with the lumber. Operators need reliable methods for making shutdown decisions while the kiln is running.

Over the years I borrowed or developed many methods that I continue to use today. Since all kilns are different, it is no wonder that the signs you are looking for are different with each kiln. It is up to the operator to find the combination of methods that gives him the best results. Here is a listing of the various methods you might try:

1. Shutdown on wetbulb level is reliable for many operators, while others find it worthless. This is truly dependent on the individual kiln. It tends to work better in steam kilns rather than direct-fired, although direct-fired kilns may find success if they consider fan direction in their decision. Kilns that share common walls cannot use this method. Readings are easily thrown off by encrusted wetbulb socks, so plan to change socks regularly.

2. The amount of steam escaping around the doors may be an indicator. Carefully feel the airflow at the base of the door. If it is too hot to touch there is still too much moisture in the kiln.
3. If your vents remain open toward the end of the charge check the color of the exhaust. Many operators pull their kilns when they see a blue haze. I think this is too late. Again, you may judge dryness by carefully touching the vent exhaust.
4. If you have a steam flowmeter on each kiln note how much steam is being consumed just before shutdown. Totalizing steam consumption through the run is also a good indicator. Both methods require that you know how much lumber is in the kiln so you can estimate consumption per board foot.
5. Direct-fired kilns can use the temperature difference between the supply and return duct. Most direct-fired systems also totalize fuel flow, although the accuracy of this value is dependent on uniform fuel moisture content.

Once the kiln is opened we have a few more indicators at our disposal:

1. If water condenses above the doors and rains down when the door is opened the lumber may not be dry.
2. We usually notice the color of the lumber first. Compare it to the color shadings of lumber you found dry in the past. Are ragged edges charred? How much in comparison?
3. Look at the exposed ends of the lumber next to the doors. We are looking for two things. First, are there tiny end checks? If you can see these there is a strong likelihood that the lumber is wet. Second, is the oozed resin sticky? The resin on dry lumber crumbles to the touch.
4. Open the doors and vents and walk down the plenum chamber between the access doors. Dry lumber has a distinctive odor. I look for my glasses to fog. How long it takes them to clear is my indicator.
5. Meter checks with a hand-held resistance meter are mandatory for a final go/no go decision. Some mills check the exposed ends next to the doors, but this limits you to only these units. Other mills push the lumber out, separate each truck, and check both ends of each unit. This gives the best sample size but is time consuming, particularly if the lumber must be redried. The most common method combines end checks with checks along the sides. Be warned that side checks do not tell you everything about the center of the unit.

Figure 12.4. After the doors are opened is usually the first real chance an operator has to judge how well he is doing his job. When doing your first checks note lumber color, odor, appearance, and texture, as well as the hot moisture content readings.

The editors want me to insert a word of caution here:

WARNING: Placing your unprotected hand in the vapor flow around the doors and vents may cause scalding. This practice is not advised and must be approached with extreme caution and only then by individuals familiar with and willing to accept the personal risks. The same cautions apply to entering a kiln soon after it is shut down. Keep low, and be prepared for a quick exit if the humidity is too high. While I may do these types of checks, no one has confirmed my sanity.

Determining the dryness indicators for your kilns is a trial-and-error job. For a new kiln, or to determine dryness indicators for an old one, start by looking at as many of these factors as possible. Keep a log of what you experienced with each load and compare this information with moisture data from the planing mill. A pattern should surface, although it will only narrow your chances for error. It does not eliminate them. Tightly constructed kilns may find many of these techniques difficult to apply. However, it pays to try.

As interest has grown toward improving the kiln-drying operation, several alternative methods for determining when to pull the kiln have emerged on the market. Each has its merits and detractors. None have proven 100 percent reliable, although this accuracy can be approached by combining several methods.

Temperature Drop Across the Load

TDAL was introduced to high-temperature drying with the first computerized multizone drykilns built in Franklin, Virginia. The concept had been around for a long time but not applied successfully to commercial kilns until then. TDAL measures both the entering and exiting air temperatures to determine the amount of heat lost as air passes through the lumber. This value is charted and is directly related to lumber moisture content. It is also a measure of the heat required to maintain drying rates.

Figure 12.5 shows how TDAL changes through a charge. After the initial high heat demand of startup, TDAL levels off at a lower level until the vents open at 8 hours. Heat demand rises to around 30° F until the fiber saturation point is passed at 12 hours. It then begins a gradual linear drop to the end of the charge. With TDAL as an integral part of a computer system, the kiln shuts down when a preset value is reached. TDAL is an average method to determine moisture content for the entire kiln by indirectly measuring hundreds of boards.

Be very careful when applying the TDAL concept to your kilns. I first learned of TDAL as an SPIB inspector in 1978. We inserted dial thermometers through holes spaced along the outside walls. A few mills installed

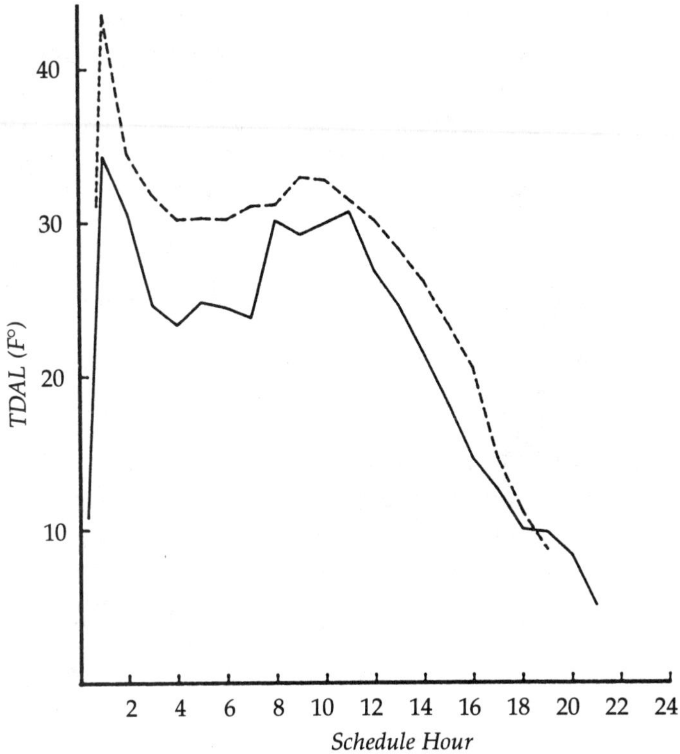

Figure 12.5. Variation in TDAL through a drying cycle. The solid line is taken from data generated from the original study done to build the first computerized multizone drykilns. Note how the heat demand drops off between the third and eighth hour. The dashed line is data taken from a recent C-M/Z installation. By venting a little during the first 8 hours we can raise heat demand and speed drying.

thermocouples along the inside walls and ran leads to the control room. They read temperatures at each location by turning a selector switch.

Neither method actually reads TDAL when installed in a two-track kiln, although it will work in single tracks. Imagine an end view of a two-track kiln with the air circulating counter-clockwise (see Figure 12.2). The temperature in the left plenum is the entering air temperature coming off the overhead heat source. The temperature in the right plenum is the cooler air after passing through two tracks of lumber and an additional heat source. Using this sensing method, there is no way to determine how much energy has been contributed by the second heat source. This value for TDAL is not an accurate estimate of TDAL because it ignores this heat source.

A well-designed TDAL monitor uses multiple temperature probes arranged on both sides of the two tracks. They should be near the lumber, not out against the outside walls or close to the center coils. This increases the likelihood of error. It is very important that stacking and baffling practices are close to perfect. Any air that circumvents around the lumber impacts on the accuracy of TDAL. As with exiting air heat control, gaps between units, missing baffles, and mixed-length lumber affect its reliable performance. Using TDAL reliably requires that the kilns perform at peak efficiency.

When making TDAL readings you must know where you are in relation to fan reversals. Fan reversals cause kiln temperatures to do a complete flip-flop, driving TDAL to zero and back up. Allow the temperatures to stabilize after fan reversals. Another consideration is precise temperature control. Oscillations in temperature around setpoint generate similar oscillations in the TDAL. Fortunately, a well-designed computerized system takes these factors into consideration, and more.

Understand the limits of TDAL. It is not 100 percent reliable, nothing is. I place its accuracy at 90 percent to 95 percent. TDAL will occasionally be wrong, even in well-run systems. It is not useful for adjusting drying rates within the individual zones, since its value is closely tied to air velocity and control accuracy. Dropping setpoints in one zone to slow drying rate affects the stability of the total TDAL reading. With air velocity varying so widely top to bottom and along the kiln length, it would be necessary to determine the optimum TDAL for each section of the kiln. A difficult task.

Resistance Probes

Resistance probes work like the familiar hammer-type resistance meters kiln operators use to check lumber. Two probes are driven into predrilled holes at precise distances from each other. Electrical leads attach to these probes and extend back to the control room. These meters are reliable only below fiber saturation point. Some considerations for their use:

1. The accuracy of these meters is entirely dependent on board selection and position within the kiln load. With the limited number of sample points these machines offer, selection is very important. Unless significant planning is done at the stacker, only edge boards may be sampled.
2. Be careful when routing wires. High-temperature kilns are notorious for dislodging boards as drying progresses. Falling boards have the potential to reduce your sensor numbers still further.
3. In high-temperature kilns the probes act as small heat sinks, increasing heat transfer in their vicinity. This undoubtably affects local moisture content although there may be a correlation between reading and actual MC.

Electrical Capacitance/Impedence Meters

Electrical capacitance/impedence meters measure the electrical capacitance or impedence between aluminum strips inserted into a lumber stack and the grounded track. The value it produces is related to moisture content. According to W. L. James and R. S. Boone of the Forest Products Lab, the following factors, among many, impact on their accuracy.

1. Readings are strongly influenced by the moisture content of individual boards the strip contacts.
2. Readings are strongly influenced by the height of the strip from the floor and its location within the kiln.
3. Readings are strongly influenced by resin buildup on the strip.

In addition, in 1980 I conducted a thesis study to determine the reliability of a commercial unit for my employer. This meter had six probes in each kiln that were read individually or averaged as needed. The study results showed poor correlation between average meter reading and average moisture content. Each probe taken by itself, however, showed an acceptable correlation between average moisture content and the meter reading. Unfortunately, each probe had very different slopes and y-intercepts when compared to each other. This meant that while the individual readings were somewhat accurate, the average of six was meaningless. Calibration was a constant concern and a problem due to the time it took.

Acoustical Emission and Sonic Meters

Acoustical emission meters measure the frequency and severity of the small pops and cracks lumber makes as it dries. This information is used to judge drying rate and adjust the schedule. Sonic meters measure the speed of sound in individual boards, which is related to moisture content. Both methods require sensitive equipment whose reliability must be proven for high-temperature kilns.

Load Cell Systems

Load cell systems measure the weight of a total volume of lumber within the kiln and records and tracks the weight as it changes through the cycle. These systems have the potential to be very reliable indicators of moisture content for the lumber they measure. The primary factor limiting their use in high-temperature kilns is the ability of the load cells to withstand the kiln environment. This is an engineering problem only.

High Temperature Drying Schedules

There is no universal drying schedule that gives optimum results for every kiln. Kilns from different manufacturers have subtle and not so subtle design differences that impact on drying speed and the setpoints they will achieve and maintain. Regional differences in timber and the individual mill sawing practices also have an impact. Finding the right schedule for each item you produce requires continuous work and testing. Your options clearly depend on the capabilities of your equipment and your ability to accept risk. You can look to other people for ideas and some guidance but it is your responsibility as a kiln operator to implement them. The disappointments are equally important as the successes since they guide you toward the solution for your kiln.

There is always room for improvement, although something major must often happen to shake you into this realization. This was driven home to me recently at the mill where we developed the computerized, multizone drying concept. During the three years I spent in the design, installation, and operation of these kilns, I had arrived at what I considered to be the "best" schedule for them. They ran with these schedules for almost five years. Yet, when the old computer system was recently upgraded to the newest generation of hardware and software, it took the mill less than a month to develop a better schedule. A schedule that raised grade recovery an additional 2 percent and shaved 30 minutes from the drying time.

The so-called authorities do not know everything. They may have a broader general knowledge and are always useful sources of ideas, but they really do not know how they will react in your kiln. For this reason I would rather train a new kiln operator who has never seen a drykiln than try to retrain someone who already knows how it is done. This reduces the likelihood that bad habits are carried over but also introduces a fresh eye to the drykiln. Someone with limited drying experience does not know what will not work and is more willing to experiment with new ideas. Certainly there will be mistakes, but there is no growth without mistakes. A poster in a plant engineer's office said it all: "No guts, No glory."

The following section begins by giving the traditional high-temperature drying schedules for both entering and exiting air control. These are good, generally safe starting points. Following this is a discussion of the different types of approaches being tried or in use, although by its nature this listing is not complete. New ideas constantly surface. Most of them can be applied successfully only to computerized controls. This is indeed an exciting time to be in the kiln drying business. The horizons are open and the potential virtually unlimited.

Beginning with your present schedule, determine how well you are doing now. This is a good time to go over the kiln to improve baffling, balance airflow and heat distributions, and improve stacking and loading. Many times these simple steps bring large gains at little cost. Collect objective data showing within-kiln moisture variations and distributions. Get a feel for how lumber appears in the stack and on the grading chain. Do boards crook out from the sides? Are the top layers ruffled in appearance? Does the lumber appear to warp after planing? Are there roller splits or long end checks or splits? Why does lumber fall into the lower grades? Quantify as much information as you can. Subjective information based on appearance can often be documented photographically to shake your memory.

Start making small schedule changes, one at a time. If you make several changes at once it is difficult to know which of the things you did made the difference. Move wetbulb or drybulb setpoints up or down. Change the timing for the vents to open. Try some of the more innovative ideas that have come along in the last few years. In some systems normal charge-to-charge variations may be so great that results are not immediately noticeable. Still, most adjustments will have an obvious impact on appearance or drying time. If you see no impact increase the magnitude of the change until you note a response. The only way to judge if you are headed in the right direction is if you have the hard data to make informed decisions. The more information, the better. Mills running computerized kilns, and which also receive moisture data from an in-line moisture detector at the planing mill, have an advantage. Not only does the comparison data have reduced chances for operator bias, but schedule changes are 100 percent repeatable. New schedules run the same way every time they are used.

Traditional Entering Air-control Schedules

There are two primary high-temperature scheduling techniques. They differ mainly in the way wetbulbs are handled at the end of the schedule. One method allows the vents to open and remain open following a preset timing period. For these kilns simply follow the schedule up to the timing period. The second method, for which these schedules are written, controls venting when the wetbulb reaches a preset temperature after timing.

Table 12.1 ***Traditional Entering Air Schedules***

	Boards	*2″ Dimension*	*4 × s*	*6 × s*
Drybulb Setpoint Entire Schedule	235° F	235° F to 240° F	240° F	240° F
Wetbulb Setpoint While Timing	Whatever it attains with the vents closed.			
Wetbulb Timing Period in Hours	4 to 6	7 to 9	10 to 14	14 to 18
Wetbulb Setpoint After Timing	150° F	150° F	170° F	170° F
Drying Time in Hours	10 to 16	16 to 24	28 to 36	40 to 56

Most kiln operators begin looking closely at their kilns when the wetbulb falls below control setpoint or below 150° F to 160° F in kilns without wetbulb control.

Exiting Air-control Schedules

Exiting air control has become common enough to include schedules of this type. Because it is a new control method to kiln drying, the full potential of its capabilities is just being explored. The schedule guidelines in Table 12.2 have been used over the last decade with success. I will discuss variations later.

Table 12.2 *Exiting Air Schedules*

	Boards	*2" Dimension*	*4 × s*	*6 × s*
Drybulb Setpoint Entire Schedule	215° F	220° F to 230° F	230° F to 240° F	235° F to to 260° F
Wetbulb Setpoint While Timing	2–5° F below whatever its maximum level is with the vents closed.			
First Wetbulb Timing Period in Hours	4 to 6	5 to 9	8 to 14	10 to 18
Wetbulb Setpoint After Timing	140° F to 160° F	140° F to 170° F	140° F to 180° F	140° F to 180° F
Drying Time in Hours	8 to 16	14 to 22	28 to 36	40 to 56

As you can see, these schedules are based on the original entering air schedules. After the timing periods the wetbulb is ramped down to its final setpoint level at a rate that minimizes drybulb temperature loss. If your system will do this automatically, you can reduce the timing periods to the lower end and below. The lower the wetbulb setpoint at the end, the brighter the lumber tends to be; however, this depends on how soon you begin venting, and at what rate.

Innovative High-temperature Schedule Ideas

The following schedule ideas have been tried and found successful at different mills running a variety of products. All of them are running computerized multizone drykilns operating on exiting air control. Due to their complexity and the importance of repeatable timing from charge to charge, I would only attempt these schedules with computerized controls.

Ramping Down Setpoints

Some years ago I designed and tried a control scheme that ramped down drybulb setpoints from about 230° F to 220° F over the course of the schedule. I used wetbulb setpoints similar to those described above. The technique was soon abandoned because it extended drying times beyond an acceptable level. No attempt was made to evaluate lumber quality, since maximum production was then of paramount importance.

About a year ago this technique was repeated at another mill, which was completely unaware of my previous attempt. They added a twist. While ramping down drybulbs from 240° F to 220° F, they began venting at 4 to 5 hours with a quick drop to 185° F on the wetbulb. The wetbulb continued with a controlled drop to around 140° F a few hours before schedule end. This is reported to improve grade and reduce drying times slightly. Moisture distributions are said to have narrowed, yet I have no quantifying data on this. A key term here is *controlled drop.* Unless you are able to track wetbulb setpoints to the lower level, you will not achieve the same results and drying times will extend. Yet this is an interesting technique that seems to work.

Ramping Up Setpoints

Another technique that is very uncommon is to vent very heavily, using 50° F to 60° F wetbulb depressions throughout the schedule, even though we continuously struggle to drybulb setpoint. The charts I have seen of this method are remarkable in that they do not reach drybulb setpoint until the very end of the drying cycle. Drying times are in line with other schedules, on the order of 18 hours; and the lumber is said to look good, with a uniform moisture content. This technique seems to be useful for kilns with inadequate heat or airflow.

High-velocity, High-temperature Schedules

This is a relatively new field, which has developed out of research done by Fred Taylor at Mississippi State. The original studies were done in an experimental kiln capable of reaching 300° F setpoints within 1 hour of startup. Air velocity was 2,000 fpm. Drying time varies from 7 to 13 hours for 2-inch dimension (2 x 6s). Several new kilns have been built based on this research, but none have attained the air velocities required to dry at the higher elevated temperatures described in the research.

REFERENCES

1. Brown, H.P., *et al. Textbook of Wood Technology—Vols 1 and 2.* New York: McGraw-Hill, 1949.
2. James, W. L. and R. S. Boone. "Capacitive In-Kiln Wood Moisture Content Monitors: Principles of Operation and Use." *Wood Science 14,* No. 4 (April 1982).
3. Southern Pine Inspection Bureau, *Grading Rules—1977 Edition.* Pensacola, FL: SPIB, 1977.
4. Taylor, F. W. and P. H. Mitchell. "Can Kiln Drying Times be Shortened Further?"*Forest Industries* (November 1987).

CHAPTER 13

Quality Control

Quality is never an accident; it is always the result of high intention, sincere effort, intelligent direction, and skillful execution; it represents the wise choice of many alternatives.

Willa A. Foster

The Problem

Peter Koch writes that "while all of the southern pines are relatively easy to dry, natural variability within the tree complicates the process. For example, the green moisture content of sapwood is higher than that of heartwood, but sapwood dries faster than heartwood. Because of this, either one may be overdried before the other reaches desired dryness." This is only one of the many interrelated factors that affect final moisture content distributions. These factors begin in the woods and extend throughout the manufacturing, handling, and drying process.

Beginning moisture contents of southern pine are related to:

- species (loblolly, longleaf, shortleaf, slash, pond, Virginia, spruce)
- tree age
- wood density
- geographical location (highland versus swamp)
- location of test within board (heartwood versus sapwood)
- height within tree

Manufacturing and handling methods that can affect moisture variation and drying rates are:

- fresh-felled or water-stored logs
- method of sawing (quarter sawn versus flat sawn)
- variations in thickness
- variations in width
- variations in surface quality
- length of time before stacking
- length of time held on sticks before drying
- sticker thickness, orientation, and placement

Kiln factors that affect moisture variation and drying rates are:

- condition of the kiln structure and panels
- condition and efficiency of baffles
- variations in heat distributions
- variations in air velocity
- humidity control method
- heating method (steam versus direct-fired)
- control method (exiting air versus entering air)
- kiln schedule
- drying time/shutdown method
- reliability of utilities (steam, water, air, electricity, fuel)
- use of equalizing and conditioning schedules

Finally, moisture variation is affected by the test method:

- testing hot versus cold
- where samples are taken (end, lumber side, lumber face, before/after surfacing)
- elapsed time between drying and testing
- tally method (hand held versus in-line meter)

Now, if that doesn't make you want to pack up your marbles and go home, nothing will. Drying lumber to a uniform moisture content in a high-temperature kiln is not a practical expectation. Given time, we could get pretty close. But that overlooks the primary reason we high-temperature dry in the first place: fast drying to reduce overall drying costs. Unless the drying system has been specifically designed to accommodate the extra time necessary for equalizing, conditioning, and extended cooldown, the best we can hope for is to narrow moisture distributions within an acceptable range. This means reducing the incidence of wet pieces to a point acceptable to quality control and customers without seriously degrading lumber by overdrying it and causing management concern.

Sample Moisture Distributions

We have no control over growth-related variations in moisture content. As kiln operators we can exert some influence over the manufacturing and handling variations, if someone will listen and champion our cause. However, once the lumber is delivered to the kilns we must do everything in our power to reduce those variables over which we have control. That is the purpose of this book: reducing the variables that affect final moisture content.

The sad fact is, even if we do everything, we still have the natural

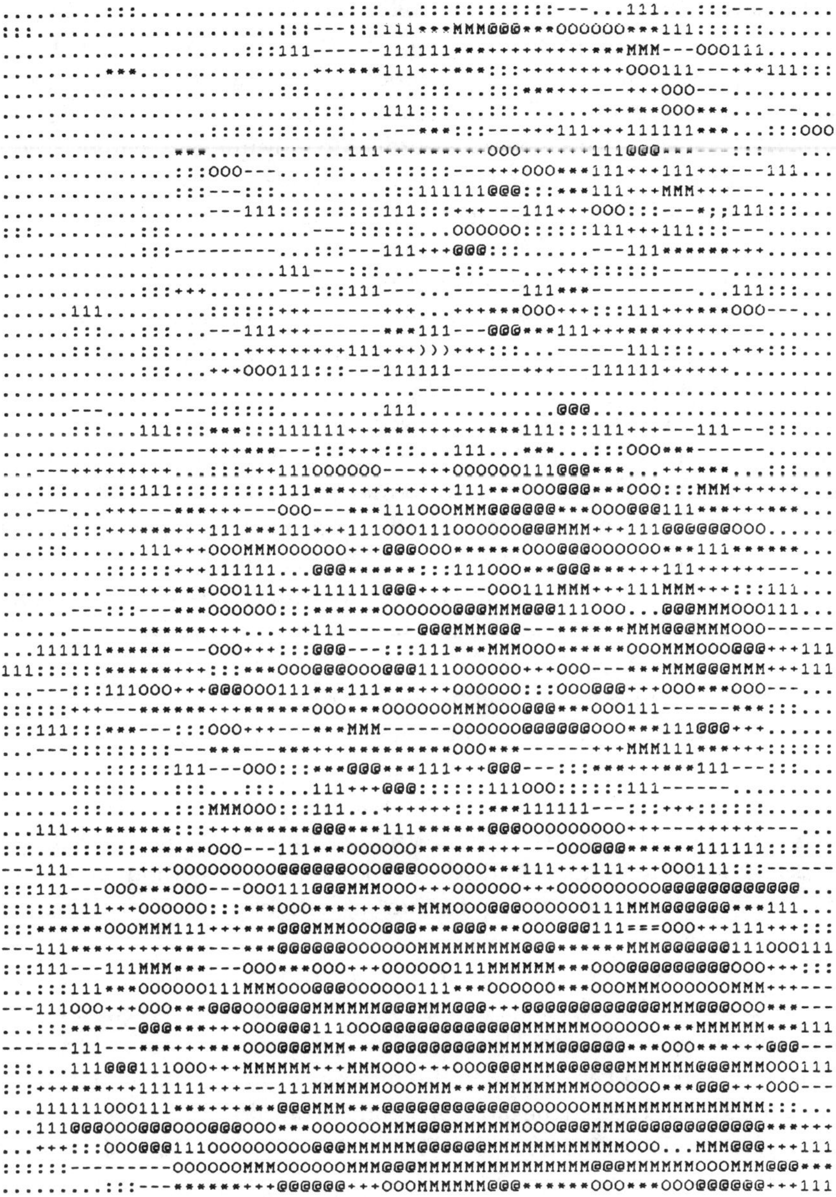

Figure 13.1. Computer-generated moisture profile of the end of a unit of lumber. Each board is plotted with a symbol assigned to indicate its moisture content. Darker shadings indicate higher moisture content.

variability of the wood and the variability inherent in any kiln design. We cannot isolate the impact of either of these factors from the other. For example, take a kiln truck of dry lumber and measure the moisture content of each piece. If we plot the position of each piece using a computer program that shades each board according to moisture content, we might get something that looks like Figure 13.1. The darker areas are the higher moisture content areas. Scattered among the dark are lighter or lower moisture content pieces. This illustrates the natural variability of the lumber versus kiln design. If moisture variation were due solely to natural variability the plot would be uniformly scattered dark and light. If the variation were due solely to kiln design the plot would be dark in the center, gradually shading to light at the edges.

We have no control over natural growth variations. But what are the variations due to kiln design? A few years back I collected detailed data from steam and direct-fired kilns to see how moisture content varied. The result of these studies were the computerized, multizone concept for steam and split return ducts for direct-fired kilns. Here is a sampling of that data:

Figure 13.2 illustrates the variation in moisture content as one travels from top to bottom in a stacked unit from each of these kilns. Both kilns were forklift loaded with two crossouts each. Although the bottom units have similar average moisture contents, the direct-fired kiln is much drier at the top. Note the dip in moisture content at each crossout. Reduce this dip by covering them. The shifting of moisture from one unit to the next is due to the variations in air velocity as we travel from top to bottom. In the steam kiln air velocity through the top unit is 950 fpm, but drops to 800 fpm in the bottom unit. Kilns that load trucks without crossouts have a gradual change from top to bottom, reflecting the gradual change in air velocity.

Figure 13.3 shows moisture variation across the width of these units. Notice how dry the edges are in both graphs, rising to the highest moisture content at the center. The direct-fired kiln has its highest moisture content slightly off center due to the differences in air velocity between fan directions. If you ran the fans constantly in one direction in either kiln, the highest moisture content would be on one side. This is why kilns that fail to reverse fan direction produce leaning packages of lumber. The high moisture content side shrinks less than the other side and packages will roll over if the difference becomes great. No wonder reversing fans is so important. If you want to center the maximum moisture contents within the units of your direct-fired kiln, run the fans a little longer in the direction blowing away from the return air duct.

Figure 13.4 presents the moisture distribution graphs for this data

STEAM-HEATED KILN
MC Variation Top to Bottom

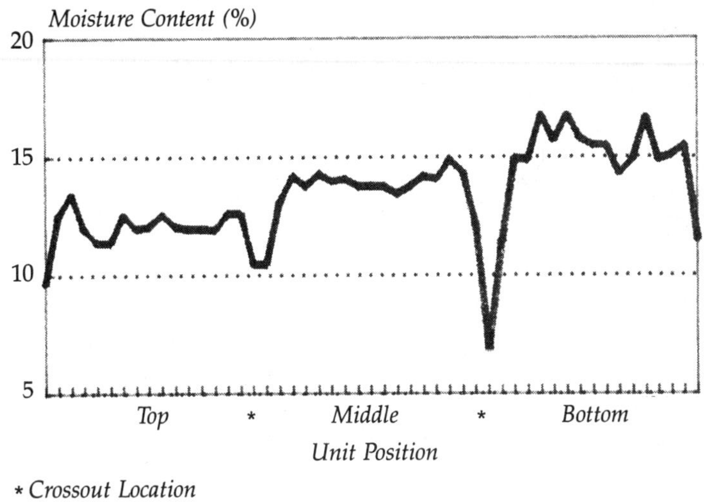

DIRECT-FIRED KILN
MC Variation Top to Bottom

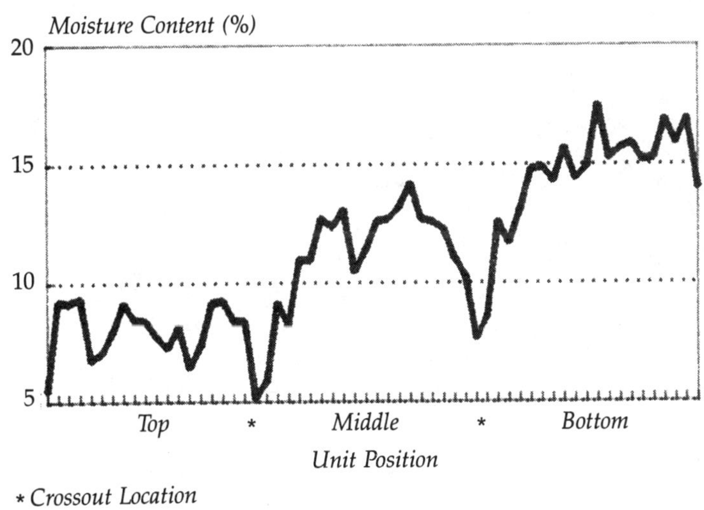

Figure 13.2. Comparison of the variation in moisture content from a two-zone steam kiln and its direct-fired counterpart. This graph illustrates the variation top to bottom in one truckload from each kiln.

STEAM-HEATED KILN
Variation Across Unit Width

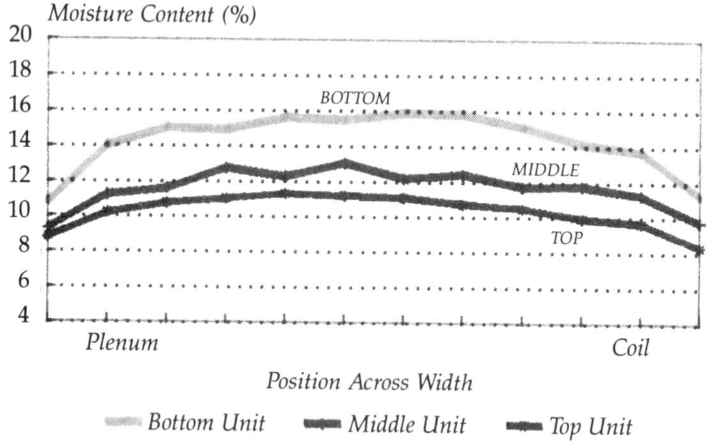

DIRECT-FIRED KILN
Variation Across Unit Width

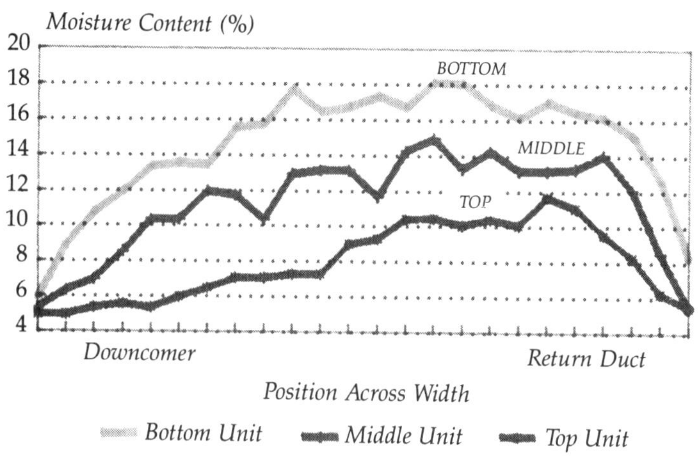

Figure 13.3. Variation in moisture content across the width of the lumber units in Figure 13.2.

MOISTURE DISTRIBUTIONS
Comparing Drying Methods

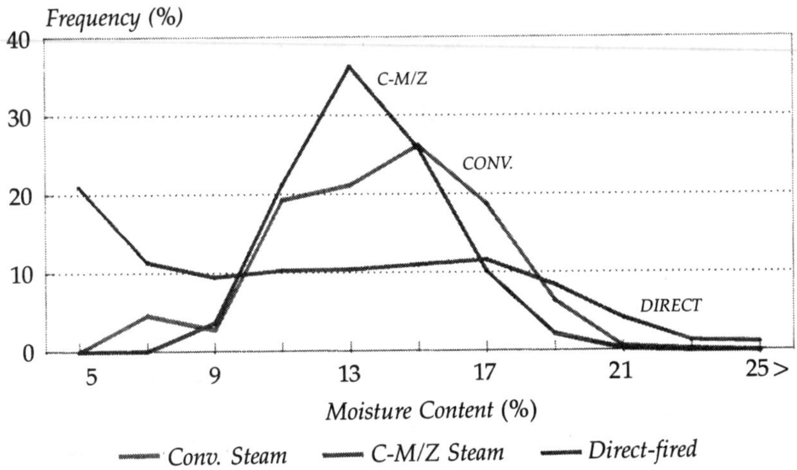

Figure 13.4. Comparison of the moisture distributions from one truckload from each of three kilns: two-zone steam, computerized multizone steam, and direct-fired.

compared to data from a computerized, multizone drykiln. All of this data was collected from kilns running standard schedules (for each kiln) without equalizing and conditioning.

Our Goal

The long-term goal of our kiln improvement program is fast, efficient drying coupled with the least degrade. Our immediate goal is to maintain the moisture content of our lumber at a level consistent with the grading rules. Southern pine mills operate under rules that limit lumber moisture content for all items of production. In addition, individual customers often have special requirements for the lumber they purchase.

Besides moisture content, our short-term goals may be different. If we are starting up a new drying system our first concern is that the equip-

Table 13.1 ***Maximum Moisture Content at Each Grademark Designation****

Item	*KD*	*KD-19*	*S-DRY*	*S-GRN*
Grade D & Btr and above, nominal thickness:				
up to 1 1/4″	90% @ 12%MC, rest @ 15%MC	N/A	15%	N/A
1 1/2″ to 2″	15%	N/A	18%	N/A
over 2″ up to 4″	15%	N/A	19%	N/A
Above 4″	18%	N/A	20%	N/A
Paneling	12%			
Board and Dimension grades, nominal thickness: (Select Structural, No. 1, No. 2, No. 3, Construction, Standard, Utility, and Stud grades)				
up to 2″	15%	19%	19%	N/A
over 2″ up to 4″	15%	19%	19%	no limit
over 4″	20%		23%	no limit
Paneling, 1″ nominal only	12%			

* Courtesy of the Southern Pine Inspection Bureau, 1977 Grading Rules.

ment functions properly. Although we can check each part of the kiln during erection, there is always a breaking-in period as we test the kiln in operation, discovering and eliminating problems. If startup was delayed (it usually is), then we may have a large volume of lumber waiting to be dried. There is always pressure to get this lumber out and speed becomes the determining factor. Just make certain the lumber is dry.

The same goes for mills that experience extended kiln downtime or sudden surges in sawmill production. Faced with mounting green inventory, quality concerns take a backseat to production.

As long as everyone understands that grade may suffer, the operator might raise drybulb or lower wetbulb temperatures to get through the crisis. If the mill is always in a crisis situation, then serious steps must be taken to increase kiln capacity. You might ignore quality in the short term, but always remember that the long-term survival of your mill depends on it.

Gaining control over your drying operation takes a lot of time and it pays to establish a realistic time frame to complete the job. Some argue that the job is never complete. There is always something else we can do to improve grade, decrease drying time, or just make things run a bit smoother. That is a good attitude to take and tends to make things more interesting in a field some find less than exciting.

The kiln operator cannot do the job alone. He needs the cooperation of everyone from the sawmill to the planing mill to management. So many times I enter a sawmill to find it divided into several seemingly autonomous areas. Each one has its own limited area of concern, to the total exclusion of others. High-quality lumber is everyone's concern, since each step in the process is interrelated to the others. An observant and cooperative approach by everyone makes the job easier for all. The benefits are shared. Stress is reduced. We spend a substantial portion of our adult lives working. What a shame if we do not enjoy it.

How do we begin your kiln improvement program? You already have. By reading this book you have begun to gather the tools required to do the job. Now apply them. Walk through your mill to identify the items that are causing you problems with drying. Look at the stacked lumber ready for the kiln. Is it uniform in dimension? Are there distortions from poor stacking and setups? What is causing them? Is the kiln operating at peak performance? How can it be improved? By paying close attention to the details it is often possible to raise grade recovery by several percentage points, usually at little or no expense. All you have to do is change the way you are doing things. To achieve maximum quality you must gain and maintain control.

As you work on problem areas up to and including the kilns you need some way to monitor your progress. How else will you know if your efforts have had any impact? Simply tracking daily grade recovery and making frequent moisture checks at the planing mill helps. The more reliable data you can get from the planing mill concerning grade and moisture the easier it is to maintain control over your drying operation. This is an ongoing effort. Once the kilns are balanced and doing their best things will still go wrong. Valves stick, traps freeze, fans go out, and wet-bulbs lose water. The list goes on. The planer mill says the lumber is wet. Is all the lumber wet, or is it just a few units? Where are those units coming from? The answers you receive tunes your response by allowing you to hunt them with a rifle rather than a shotgun. Kiln operators who arbitrarily add or subtract time from their schedules in reaction to complaints of wet or overdry lumber from the planing mill are out of control. They are doomed to swinging from one extreme to the other. How can they possibly relax?

Record Keeping

The more information you keep on your kilns and the lumber they produce, the better. This gives you a solid base of information for future decisions. For example, I worked at a mill that ran two kilns with internal direct-drive fans. We were losing a lot of fan motors. Because we kept records of our losses we were able to determine how much it was costing us and justify replacement with external fan drives. On a daily basis complete information helps you track down the reason for drying problems related to grade, moisture, and drying time.

Computerized controls that generate and store important process data for later retrieval help. I was once called into a mill to investigate the reason for extended drying times. We checked over all the systems in the kilns and could find only minor items that would not account for the extra time. The computer system on the kilns tracked and stored the usual drybulb, wetbulb, and moisture content data, but also tracked steam pressure. By plotting out the kiln charts over the previous month or so and laying them out on the floor, we were able to determine that something had happened at the boiler. This type of problem usually goes undetected.

I am often reminded of a kiln operator I once worked with who had been drying lumber for a long time. Since I was new to this business I often went to Willie to ask his advice about peculiar drying problems I was seeing at the planing mill or drykilns. He would look thoughtful and slowly rub his chin. "Well, you know," he would say, "that always happens this time of the year." It didn't take me long to start looking for my own answers and I found that most of the problems I was seeing were not seasonal. Something else was going on. The only way I could determine what was by keeping detailed records.

Following is a listing of the type of information you will find useful.

General information:

1. Service dates for preventive maintenance.
2. Information on emergency repairs, including dates, nature of the repair, and individual component (identify trap, fan, motor, bearing, etc.).
3. Interruptions of power, water, heat, or air.

Individual kiln charge information:

1. Items dried (2×4s, 2×10s, 4×4s, etc.).
2. Lumber location within the kiln (tag lumber for later identification).
3. General quality of the green lumber (fresh, molded, etc.).
4. Schedule.

5. Drying time.
6. Turnaround time.
7. Final wetbulb reading at shutdown.
8. Final reading of any meter used to determine when to pull the kiln.
9. General appearance.
10. Information from method used to determine dryness at the kiln (hot meter readings, etc.).
11. Information from the planing mill on grade and moisture content.
12. Kiln charts and printouts. Steam pressure and demand charts from the boiler.
13. Detected problems with the charge.

The more information you can routinely gather from this list, the easier it is to run your kilns. Try to design a standard form to fill in with the charge information and keep on file for each kiln for at least a year. Sooner or later you will need it.

The rest of this section discusses methods for gathering and organizing moisture data from the kiln and planing mill.

Tracking Lumber Through the Kiln

Item 2, above, recommends tagging each unit of lumber so you can track it through the kilns to the planing mill. This is a very useful tool for isolating drying problems. If your mill uses inventory tags, then these will do. Simply keep a sheet recording tag numbers and location within the kiln. Some mills print tags specifically for this purpose and include important information on them. Printed on heavy index card stock, they may brown but I have never known one to catch fire inside the kiln. Use standard ¼-inch staples to attach them at the stacker or before loading into the kiln.

Several mills use tags perforated into three sections, with each section carrying the inventory number for that unit. As the lumber passes from the stacker to the planing mill, sections are detached to keep inventory up to date. At the stacker the first section is detached with a record of lumber size and stacking date. At the kilns the second tag is removed to track drying date and serves to locate the units within the kiln. Finally, the rest of the tag is removed at the planing mill. Information is added here on quality and moisture content. This allows you to track lumber throughout the system, accumulating much useful information along the way. This entire process is becoming increasingly automated, with tags designed to be read by barcoders tied into a computer system. These systems are primarily used to keep up-to-date inventories, but also allow you to stage lumber into the kilns on a first-in, first-out basis.

Figure 13.5. Lumber tags are useful for tracking lumber from the kilns to the planing mill. If high moisture is found the tag is removed and sent back to the kiln for action. Photo courtesy of Union Camp Corporation.

Whatever system you use, try to spend a little time at the planing mill each day pulling tags and checking moisture content. Your concern helps foster the cooperation you need to keep the system running smoothly.

Collecting Moisture Information at the Kiln

Your first chance to judge the quality of a charge occurs when you open the doors of the kiln. Up to that point you have relied on the color or volume of vapor coming from the vents and doors to tell you the lumber is dry. You might monitor wetbulb or have an in-kiln meter to judge dryness. Once inside, however, you can look at the color of the wood and breathe the vapors that tell the experienced eye and nose that the lumber is dry. Even so, it is only an educated opinion. It is time to bring out the moisture meter.

Use a resistance meter with long, insulated needles. These meters are available with southern pine dials that have timber species correction built in. This is important if you plan to use the meter in the planing mill as well as at the kiln. I have never found meters with built-in temperature

Figure 13.6. Check your lumber at the kilns to head off trouble. Photo courtesy Union Camp Corporation.

or species correction very useful at the kiln. They only seem to add to the confusion of running them. We use our meter to establish a correlation between moisture checks at the kiln and those at the planing mill. Compensation for temperature is not necessary for this. I still recall the look of exasperation on the face of a kiln operator who had been told to subtract 25 percent from his hot meter readings to get the lumber moisture content. "How," he exclaimed, "do you subtract 25 percent from a meter that doesn't read!" That is a very good question.

There is equipment that allows you to penetrate to the inside of the unit to measure moisture. The most common of these, known as a "T-bar," is simply another electrode for your resistance meter. Their shallow penetration depth limits its use, particularly in lumber that has a wide moisture gradient from center to outside. Since you cannot see the board you are testing, you cannot know if it has some defect that affects the reading (pitch or knots come to mind) or if you are even in one board. You may be testing two boards at once. Some mills mark the handle of the T-bar so they can judge where the middle of the board is.

Store the meter in a box heated with a lightbulb. This keeps the electronics warm and dry to maintain accuracy. If you keep it in an air-conditioned control room and take it outside or into the kiln, water condenses on it and renders it worthless for your purpose.

Do not use short needles or needles with blocks designed to limit penetration and prevent their breakage. These only allow you to check the surface and may not give you accurate information at the kiln. They are fine for the planing mill. Always drive the needles into the lumber so they align with the wood grain. Measuring across the grain alters the accuracy of the reading. Testing across the grain is acceptable only when probing the ends of units to determine the correlation between hot and cold readings.

Calibrate the meter before each use and change the batteries frequently. Always check calibration using standard calibration blocks available from the manufacturer. They are inexpensive insurance against equipment malfunction.

How you check your lumber at the kiln is a matter of personal preference. Some mills check it within the kiln, first testing the exposed ends and then recording moisture from the sides of each unit. Other mills push the lumber out and test the sides as well as each end. This has the advantage of allowing you the ability to check each board within a unit, not just the sides. As you probe each board, write its moisture content next to the test spot with a crayon. If you later decide to run the kiln a bit longer, you can easily compare readings to see how the moisture dropped. This is extremely useful information when trying to gauge how much longer you must run a future charge that is not quite dry.

Checking the sides within the kiln is hard to beat in terms of speed.

Keep in mind, however, that the information from this method does not always reflect the true dryness of the lumber. For example, if a fan goes out during the charge the outside of a unit may check dry while the interior is wet. Still, it is a useful method as long as you understand its limitations.

Whichever method you use you must determine the correlation between the readings at the kilns and the final moisture content at the planing mill. The technique you use is important and must be standardized. In other words decide how you want to check your lumber and always check it in the same manner. Changing anything affects the reliability of the technique. If you have an established technique and change your kilns, particularly the control method (exiting versus entering air), expect your testing results to change. Unfortunately, there are no guidelines to help you anticipate what the change will be. Here are the factors you must consider when deciding on a sampling method:

1. Which boards are sampled and where? Do you check on the sides or ends? The package middle or edge? Look at the typical moisture patterns at the beginning of the chapter or break down a few units to see what you have.
2. When do we sample? In the kiln while it is still hot, or outside after a regular cooling period?
3. How deep do the needles penetrate? Do we check full depth or use a block to prevent breakage?
4. How many boards do we sample? Do we ignore or not sample certain ones?
5. What results determine if we accept or reject individual units or even the entire charge?

Changing any component of the test procedure changes the results and accuracy of our testing method. Be consistent in your testing and record your results. Over time you tune your technique. For example, some mills check the outside of their units within the kiln. They drive the needles full depth, avoiding heavily pitched or wide-grained pieces. They have found that they are in good shape as long as they get no readings over 18 percent. One or two percent above this and they will put a little more time on it. They know from experience how long that will be. Our mill, on the other hand, tested both unit ends outside the kiln after the lumber had cooled for 15 minutes. Heavy pitch was ignored. We looked for the majority of our readings to be around 15 percent. Combined with a technique known as acceptance sampling we reduce the risk that wet lumber gets to the planing mill.

Acceptance Sampling

The moisture content of our lumber is not uniform but varies within a normal distribution. Any moisture checks we make also vary within this distribution. Acceptance sampling attempts to determine the moisture distribution of our lumber by checking only a few boards. In actuality, this is what we do when we sample the outside of units as we decide when to pull the kiln. This is a more formal approach, but keep in mind that "No sampling plan can give complete protection against the acceptance of a defective product."[1] No matter what you do, some wet lumber will reach the planing mill. Our purpose is to keep it within acceptable limits.

Sampling must be consistent. Develop a technique and stick to it. Any changes will have unpredictable results. Finally, keep historical data on the sampling procedure to evaluate its accuracy over time.

Using the factors mentioned above for establishing a sampling method, begin collecting hot moisture data from individual units at the drykiln. Record the hot readings from 10 to 20 boards from each unit and compare this information to the total moisture distributions of each unit at the planing mill. Look for a trend.

We developed the following procedure at the mill that tested lumber after it was pushed out of the kiln. We probed 10 boards in each unit, driving the needles full depth into the end. We chose boards at random, although we excluded the outside two layers from consideration. We knew these boards always tested dry. If any board read 18 percent or higher we counted it. If three boards out of ten read 18 percent or higher we held that unit for redry. If more than four units in a charge were held for redry the entire charge was loaded back into the kiln and more time added. Figure 13.7 shows sample distributions from this testing method.

After a while we found that we really only needed to check the bottom units because that was where the highest moisture was found. If a bottom unit was wet, then we checked the unit above it. In this way, we reduced the possibility of wet lumber making it to the planing mill. Each mill is different and must establish its own sampling procedure. This requires a great deal of work and commitment, but the rewards are there.

Do not use the sampling procedure to reduce the frequency of wet units to zero. I have moisture distribution data from virtually every unit within a single high-temperature kiln charge. What is remarkable about this data is that the outlines of each unit's moisture distributions are identical, even though their average moisture contents vary. This demonstrates that if we run our kilns long enough to reduce the number of rejected units to zero, we do not change their distributions. We merely overdry increasing amounts of lumber.

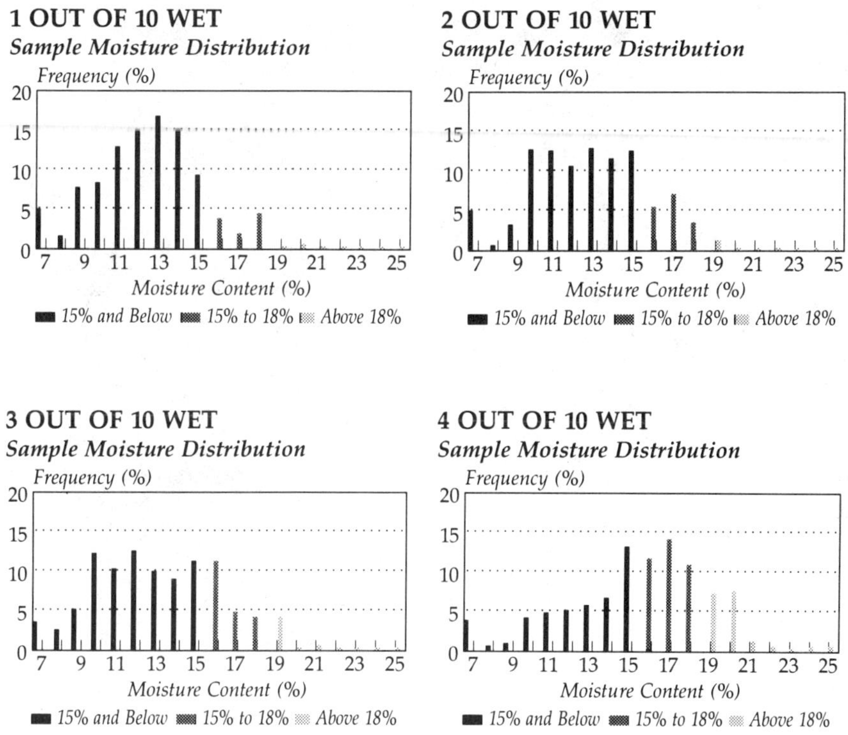

Figure 13.7. Sample moisture distributions obtained while developing an acceptance sampling technique. One out of 10 wet means that of 10 boards from 1 unit sampled hot at the kiln, 1 read 18 percent or higher.

Monitoring Moisture Content at the Planing Mill

Unless you are collecting moisture content information from the planing mill, you are leaving too much to chance. Many mills assign the grader to check moisture content. Although they are the logical ones to do this since they see all the lumber, the probability of bias in the data is great. Graders often check boards only when they suspect wet lumber or to confirm that the lumber is "burned up." This information, although it serves the graders' purposes, is worthless unless you can determine the kiln charge and location of the lumber within that charge. It is not representative of the lumber. Dependence on this data-collecting method is very risky, although few mills have any alternative. Even if you collect moisture data yourself, you do not see all of the lumber. There is also a tendency toward bias in the other direction.

Figure 13.8. Check moisture periodically on the planing mill breakdown.

How do you eliminate bias errors from checking lumber by hand? You cannot, although you can reduce it by limiting the ability to arbitrarily select lumber for testing. One method is to preselect boards by spray-painting a diagonal stripe on the inspection ends of your units. Start in the upper right corner and spray a line to the lower left. On the grading chain only the painted boards are tested and recorded. Have the graders check ten or more boards an hour. If this is too tedious set up a timer with a light that comes on at preset intervals and hang it in the grader's view. When the light comes on he begins checking boards at random until it goes out. A few minutes once an hour is plenty. By the end of the day you will have data from several hundred boards.

The handiest tool for maintaining quality control over the dry kiln is the in-line moisture detector. These machines sample and tally each board automatically, eliminating the bias errors of human intervention. Most detectors allow you to sample individual units to compare lumber from different areas of the kiln or between kilns. I would not attempt to run a high-production operation without one. However, reports are beginning to emerge concerning a tendency for this equipment to drift, or lose cali-

Figure 13.9. The graders are valuable allies for improving your drying operation. Photo courtesy Union Camp Corporation.

bration. Several mills report that maintaining calibration is an ongoing struggle and that it must be checked daily. Now that this problem has been identified, I expect that steps will be taken to correct it in future models.

Planer operations with computerized trimmers and sorters produce detailed information on grade recovery and trim. This information is difficult to compare from one charge to the next and is not suitable for this purpose. However, it will tell you how you are progressing with your kiln improvement program over long periods of time.

Applying Quality Information to the Kiln

The first rule is: Never overreact. When the planer supervisor starts screaming about wet or overdry lumber, look at your information. If you have kept good records it will be easy for you to track down the solution. Was this an isolated unit or was the entire charge wet? Was there an equipment malfunction? Be very reluctant to make drastic changes in your schedules. After all, they worked OK before.

Figure 13.10. How your lumber appears on the grading chain tells a lot about how good a job you are doing with your drying program.

Examine the information available to you. Get the planer personnel involved in determining the extent of the problem. Finally, make an informed decision.

REFERENCES

1. Grant, E. L. *Statistical Quality Control*. New York: McGraw-Hill, 1952.
2. Koch, Peter. *Utilization of the Southern Pines. Vols. 1 and 2*, Agricultural Handbook No. 420, Forest Service, Supt. of Documents, GPO Stock No. 0100-1483, 1972.
3. Southern Pine Inspection Bureau, *Grading Rules—1977 Edition*. Pensacola, FL: SPIB, 1977.

CHAPTER 14

Kiln Maintenance and Design

To love what you do and feel that it matters—how could anything be more fun?

Katharine Graham

Preventive Maintenance

An important part of running a successful drying operation is understanding the drying process well enough to recognize when something has gone wrong. Unless you are very observant of things that are out of place, the best way to accomplish this is by establishing a preventive maintenance program. Rather than react to sudden equipment failures, periodically check each system component and make routine repairs as required. It is much easier to schedule downtime than to catch up following unexpected emergency repairs. Worse, some problems are slow, progressive failures, which take some time to detect. Head them off by checking each part of the kiln on a regular basis.

The checklists at the end of the chapter are designed to walk you through the kilns, checking the important components that assure reliable operation. One checklist covers steam-heated kilns, the other direct-fired. Modify one to suit your specific needs.

Reading Kiln Charts

Drybulb

Symptom	*Possible Cause*
Drybulb above setpoint	1. Control valve stuck open 2. Control valve bleeding through 3. Heat entering from attached kiln 4. Bypass valve left on

Drybulb below setpoint	1. Control valve frozen closed 2. Hole in control valve diaphragm or air line 3. Vents stuck open 4. Loss of steam pressure 5. Loss of air pressure 6. Controller malfunction 7. Traps not working 8. High boiler backpressure
Drybulb shifts with fan reversals	1. One vent row is stuck closed. 2. Drybulb sensor positions are not symmetric from one side of the kiln to the other. 3. Poor baffles or gaps between the lumber units allows air to escape around lumber (exiting air systems). 4. Very different airflow from one fan direction to the other. 5. Sensor calibration is wrong on one bulb. 6. Only one sensor is being read due to a broken relay. 7. One capillary is crushed on a dual-bulb vapor-bulb controller. 8. In multizone systems, one bulb may be out.
Drybulb oscillates about setpoint	1. Heat radiation on the temperature sensor causing excessive cycling. 2. Controller needs tuning.

Wetbulb

Wetbulb above setpoint	1. Vents are not functioning, check: a. diaphragm or positioner b. for leaks in air line c. malfunctioning controller 2. Large steam leak. 3. Sock becoming encrusted or drying out from lack of water. 4. Steam sprays on. 5. Insufficient venting area.

Wetbulb below setpoint	1. Vents stuck or blown open. 2. Insufficient moisture in kiln to maintain setpoint. 3. Wetbulb laying in water.
Wetbulb shifts with fan reversals or otherwise	1. One vent row is not coming open. 2. Moisture seeping in from adjacent kiln. 3. Steam leak. 4. Sock is encrusting.
Wetbulb oscillates around setpoint	1. Too much vent area. 2. Controls need tuning.

Fire Prevention

The following is excerpted from "Fire in the Dry Kiln" by Kent Townsend:

> Fire in a dry kiln can be a disastrous occurrence. Although there are some reported incidents of fires in steam kilns most fires occur in direct fired kilns and most often in those kilns using a wood fired suspension burner.
>
> Fires that result from wood fueled suspension burners are not that infrequent, especially on poorly cleaned and maintained burners. Mills that have these types of kilns and burners soon realize that close observation of the kiln during operation is necessary in order to detect fires quickly and take prompt action. The importance of periodic cleaning of accumulated material in the burner, adjustment of the fuel/air ratio to manufacturers' specifications, and the repair of refractory and general burner equipment are all very critical to the prevention of fires in the dry kiln.
>
> Fires seem to begin from two major causes: 1) uncombusted fuel which causes sparking, 2) leaks in the air distribution duct.
>
> The most successful way to (fight a fire in the kiln) is to first be prepared and have a plan before the fire ever occurs. Discuss the plan with the kiln people and/or those who will fight the fire. When the fire is discovered, shut down the kiln including fans, burner, vents and doors, sound the alarm, call for assistance, and activate the sprinkler if not already activated. Once assistance arrives prepare the fire hoses and fire system to extinguish the fire once it is outside the kiln. Once everything and everyone is ready, push the load out towards the green end with a forklift. Extinguish the fire. Some common sense regarding the size of the fire should be exercised before attempting to push it out. If it is too

> large it could endanger the life and safety of the fork lift driver in which case the fire will have to be fought in the kiln. Most fires of this nature, however, give ample time for careful thought and action if the kiln itself is not burning.

I might add that sprinklers should be dry pipe rather than wet pipe systems. Wet pipe systems tend to deteriorate quickly in the kiln environment.

Design Tips for Easier Kiln Operation

Part of my business is designing kilns and drying systems that accomplish their intended purpose. Not only must they be capable of drying high-quality lumber, they must also be easy to use. With a little thought you can avoid many of the chronic problems that plague drykiln operations. Take fan maintenance as an example. If maintenance personnel must squeeze through small access doors or clamber over or through pipes and fans in a hot environment: (1) They will be reluctant to do their jobs because of the physical discomfort; and (2) they will likely rush through their work and miss something. Give them easy access, cool working conditions, and plenty of light and there is a better chance that the work will be done right. It may cost a little more up front, but the reward is a drying system that operates at higher efficiencies. This translates into dollars over the long haul.

Foundations and Track

Raise the perimeter foundations on which the columns rest a few inches off the floor. Since water pools on the floor in the early stages of drying, there is a tendency for the bases of these columns to rust out if built on a flat slab. Supply drainage inside of the kiln to remove excess water.

Bury tracks in concrete up to the ball. This exposes less track to corrosion and is a lot easier to work and walk around. Install lift-out rail sections under the doors, but make certain the foundations between the kilns and loading aprons are properly aligned and will remain so. Without these lift-out sections, expect track erosion under the doors.

Baffles

Use hinged baffles for the floors, overhead, and ends constructed of heavy 16-gauge steel. The lift cables on the overhead baffles must lift straight up, not out, to reduce the effort required to raise them. Concrete floor baffles do not seal well enough and should be avoided. Use chains rather than springs to hold the end baffles in place. No baffles should impede the flow of air. They must contact the lumber stacks at their edges.

Panel Design and Insulation

Panels must be designed to allow the escape of moisture, which enters and condenses inside. It must also prevent shifting or sagging of insulation material. Pay close attention to the panels on new kilns to detect bulges or hot spots on the outside. This is an indication that problems are beginning. Bring them to the attention of the manufacturer for resolution.

There is a continuing controversy over the merits of different insulation materials, panel thickness, and R-values. Let's talk about R-values first.

R-value is the resistance a material has to heat transfer. It is expressed by the equation:

$$\text{Heat Loss (BTU/HR)} = \frac{\text{Temperature Difference} \times \text{surface area}}{\text{R-Value}}$$

Let's look at an example comparing the heat loss from a double track by 68-foot high-temperature drykiln with R-9 and R-20 insulation. A kiln of this size will have approximately 6,936 square feet of surface area and hold 100 mbf of lumber. Assume the kiln interior temperature is 240° F with an outside temperature of 65° F for a total temperature difference of 175° F.

For R-9:

$$\text{Heat Loss} = \frac{175^\circ \text{ F} \times 6{,}936 \text{ sqft}}{9} = 134{,}867 \text{ BTU/HR}$$

For R-20:

$$\text{Heat Loss} = \frac{175^\circ \text{ F} \times 6{,}936 \text{ sqft}}{20} = 60{,}690 \text{ BTU/HR}$$

Heat loss difference between R-9 and R-20 panels = 74,177 BTU/HR

or,

Total pounds of Steam per Hour @ 870 BTU/LB = 85.3 LB/HR

Average steam demand required to dry 100 Mbf of lumber in 20 hours:

$$\text{Average Steam Demand} = \frac{100{,}000 \text{ bf} \times 2.8 \text{ \# steam/bf}}{20 \text{ hours}} = 14{,}000 \text{ LB/HR}$$

Heat loss through the panels as a percentage of average steam demand $= \dfrac{85.3\text{ LB/HR}}{14{,}000\text{ LB/HR}} \times 100 = .61\%$

In terms of the total heat consumption of a drykiln, heat loss through the kiln skin is very low. If a mill is burning woodwaste for fuel, it is insignificant. A more lasting concern is the long-term durability of the insulating material. After all, we do need insulation in our kilns. Rather than enter into a potentially litigious controversy over insulating materials, I will simply list what is available and give you some guidelines for selection. I have seen them all and have my own opinion. Let the salesmen slug it out.

The following materials are available:

- fiberglass board
- foamed glass
- polyurethane foam
- polyisocynurate foam
- hybrid boards combining layers of different materials
- mineral wool

The insulation you select must be durable and not retain water. It must not lose its shape in the panel. It must be chemically inert to wood chemicals under high temperatures and saturated humidity conditions. It must not degrade over the life of the kiln. Ask to see kilns with the proposed panels that have been in operation for over a year. Inspect them personally. Look for bulges or distortions in the panel. Run your hand over the surface of the panel while the kiln is running to detect hot spots. There are good panels out there, you just have to discover them.

One comment. Back when I did kiln surveys as an SPIB inspector, we often punched holes in the sidewalls to insert dial thermometers to check temperature balance. Never do this. Moisture-laden air pushes into the panels by the internal kiln pressure and its moisture condenses on the interior of the outside panel. Water soon saturates the insulation in that panel and reduces its effectiveness.

REFERENCE

1. Townsend, Kent. "Fire in The Dry Kiln." *FPRS Wood Drying News Digest*. Madison, WI: FPRS, 1981.

Preventive Maintenance for Steam Kilns

Daily Routine	
	Change wetbulb socks if they have a tendency to become hard quickly. Check water flow.
	Listen for steam leaks when you open the kiln, locate and schedule their repair according to severity.
	Check fit and action of the baffles. Schedule repair of damage.
	Check sticker alignment and kiln setups. Spot check lumber for manufacturing problems affecting your drying.
	Lubricate kiln trucks and remove damaged ones for repair.
	Remove any fallen boards from the kiln.
	Make certain the doors seal.
	Make certain control valves open on startup.
	Make certain all fans are running. Investigate the reason for those which trip out frequently.
	Check your traps if they are outside. If you have a computerized multizone kiln check your screen displays to verify that heat is coming from each coil.
	Check kiln charts for poor heat rise or too high a wetbulb and correct any problems.
	Spot check moisture content at the planing mill.
	Drain water from air filters and lines.

Weekly Routine	
	Replace wetbulb socks (unless you do it on a daily basis).
	Check traps and control valves by turning on the heat and listening to the traps with a mechanic's stethoscope.
	Sweep out the kiln to remove accumulated dust.
	Do scheduled repairs – steam leaks, baffles, and so on.
	Record fan motor amperage.

Monthly Routine	
	Grease fan motors and bearings, and inspect fans for damage. Check fan clearance and rotation. Don't forget to close access doors when finished. Check rotation and amperage. Tension and replace belts if required.
	Replace trap kits in the traps scheduled for their yearly inspection.
	Inspect kiln walls and doors for deterioration and schedule repairs.
	Inspect temperature sensor mounts for damage.
	Clean out the underground condensate line trench.
	Inspect vents and linkages.
	Grease vent shafts on vents with internal linkages.

Once a Year	
	Check fan pitch with an inclinometer.
	Inspect tracks for damage.
	Inspect the area at the base of the doors for damage and future repairs.
	Inspect steam supply main and headers for necessary insulation and steam leak repairs. Replace the kits in traps on the drip legs.
	Inspect the heating system for shifting or sagging coils.

Preventive Maintenance for Direct-fired Kilns

Daily Routine	
	Change wetbulb socks if they have a tendency to become hard quickly. Check water flow.
	Check fit and action of the baffles. Schedule repair of damage.
	Check sticker alignment and kiln setups. Spot check lumber for manufacturing problems affecting your drying.
	Lubricate kiln trucks and remove damaged ones for repair.
	Remove any fallen boards from the kiln.
	Make certain the doors seal.
	Make certain all fans are running. Investigate the reason for those that trip out frequently.
	Check kiln charts for poor heat rise or too high a wetbulb and correct any problems.
	Spot check moisture content at the planing mill.
	Drain water from air filters and lines.
	Follow manufacturers' instructions for burner and fuel handling equipment maintenance.

Weekly Routine	
	Replace wetbulb socks (unless you do it on a daily basis).
	Sweep out the kiln to remove accumulated dust.
	Wash ash from downcomers and kiln interior.
	Do scheduled repairs – duct damage, baffles, and so on.
	Follow manufacturers' instructions for burner and fuel handling equipment maintenance.
	Record fan motor, recirculation, and combustion air motor amperages.

Monthly Routine	
	Grease fan motors and bearings and inspect fans for damage. Check fan clearance and rotation. Don't forget to close access doors when finished. Check rotation and amperage. Tension and replace belts if required.
	Grease recirculation and combustion air blower.
	Check distribution and supply ducts for missing bolts and damage. Turn on the recirculation blower and check again.
	Scrape ash deposits from fan blades and recirculation blower vanes.
	Inspect kiln walls and doors for deterioration and schedule repairs.
	Inspect temperature sensor mounts for damage.
	Follow manufacturers' instructions for burner and fuel handling equipment maintenance.
	Inspect vents and linkages.

Once a Year	
	Check fan pitch with an inclinometer.
	Inspect tracks for damage.
	Inspect the area at the base of the doors for damage and future repairs.
	Inspect fuel blower line for rust and repair.
	Inspect hammermill for wear twice a year.
	Follow manufacturers' instructions for burner and fuel handling equipment maintenance.

Glossary

ACCEPTANCE SAMPLING A statistical quality control method that seeks to determine the quality of a product by sampling a small portion of it. It does not assure that 100 percent of the product is acceptable, but reduces the likelihood of acceptance of a defective product (high moisture content).

AIRFLOW The pattern of air movement within the drykiln. This pattern varies with kiln design, loading methods, and level of maintenance and operation.

AIR VELOCITY The speed of airflow through the stickers, crossouts, and around the lumber expressed in feet per minute (fpm) or meters per second (mps).

ANNUAL RING One year's growth increment of a tree composed of one band of springwood (earlywood) and one band of summerwood (latewood). Also called a growth ring.

BAFFLE A metal, concrete, or wooden barrier used to deflect and control airflow through the lumber.

BAFFLE, END A hinged baffle that closes the gap at the end of the lumber next to the doors.

BAFFLE, FLOOR A hinged or stationary baffle that closes the gap at the bottom of the lumber unit, preventing air from escaping under the load.

BAFFLE, OVERHEAD A hinged baffle that closes the gap between the overhead fan deck and the top of the lumber load.

BARK The outside layer of a tree stem. It provides protection, insulation, and reduces moisture loss.

BOILER Steam generation equipment.

BOLSTER *See* CROSSOUT.

BOUND WATER Water that is bound within the cell wall of wood. Water held in wood below the fiber saturation point. Bound water moves by diffusion.

BOW A deviation flatwise from a straight line drawn from end to end of a piece of lumber. *See also* WARP.

BULBS General term referring to the drybulb and wetbulb sensors within a kiln.

BUNK *See* CROSSOUTS.

BUNK SUPPORT The horizontal kiln truck supports on which the lumber is placed for drying.

CALIBRATION Adjusting the control or recording equipment to reflect the actual control or recording temperatures.

CAMBIUM The layer of tissue dividing the bark from the wood, which forms new bark to the outside and new wood to the inside as the tree grows.

CAPILLARY ACTION The combination of solid-liquid adhesion and surface tension by which liquid moves through a cellular structure.

CASEHARDENING A condition of stress and set in dry wood in which the outer fibers are under compressive stress and the inner fibers under tensile stress. The stresses persist after the lumber is dry and cause warp if the wood is remachined after drying. *See* REVERSE CASEHARDENING.

CHECKS A splitting of the wood fibers within or on a log or lumber. Checks result from uneven wood shrinkage.

COILS The radiant heating surface within steam-heated kilns. Steam coils contain the steam and transfer its heat through the coil wall to heat the kiln air, producing condensate in the process.

COILS, CENTER The steam coils that are positioned between the kiln tracks. Also called *booster coils*.

COILS, OVERHEAD The steam coils positioned overhead at or on the fan deck.

COMPRESSION WOOD Reaction wood in coniferous (softwood) trees, which forms on the underside of leaning trees and limbs. This wood is characterized by high longitudinal shrinkage as compared to normal wood.

COMPUTER SIMULATIONS Computer software that models actions or occurrences in the real world. By using modeling or simulation techniques on the computer, you can experiment with various drying techniques without risk of destroying lumber.

CONDENSATE Water formed by removing heat from steam within the kiln heating coils.

CONDITIONING A process for relieving the stresses present in wood at the end of drying.

CONTROL VALVE A device that controls the flow of steam to the steam coils.

CROOK A deviation edgewise from a straight line drawn end to end of a piece of lumber. *See also* WARP.

CROSSOUTS Horizontal spacers between units of lumber. They align over the sticks and serve to separate the units for forklift handling. Also called *bolsters* and *bunks*.

CUP Deviation in the face of a piece from a straight line drawn from edge to edge of a piece of lumber. *See also* WARP.

DEADPILE Stack lumber in units without stickers.

DEGRADE Drop in lumber grade due to manufacturing or drying practices.

DENSITY The weight of wood in relation to its volume. If not measured directly, density is inferred from the number of rings per inch. The more rings per inch, the higher its density.

DE-SUPERHEATER A device for removing the excess heat in steam as its pressure is reduced.

DIFFUSION Movement of moisture from areas of high to low concentration or temperature.

DOOR CARRIER A device that lifts kiln doors and moves them away from the kiln opening along an overhead track.

DOORS, ACCESS Small doors, usually located at ground level, which permit operator access to and escape from the kiln interior.

DOORS, KILN Large openings through which lumber passes when loading a kiln.

DOWNCOMERS Small distribution ducts within direct-fired kilns, which distribute heated air between the kiln tracks from an overhead main distribution duct.

DRYBULB A sensing device (thermometer) that measures the temperature of the air.

EARLYWOOD Thin-walled wood cells deposited by the cambium during the fast growth of spring. Also called *springwood*.

EDGING Trimming the width of a board of defects.

ENTERING AIR CONTROL Sensing and controlling drybulb temperatures of heated kiln air as it enters the lumber load.

EQUALIZATION The process of narrowing the moisture distribution between boards at the end of a drying charge.

EQUILIBRIUM MOISTURE CONTENT (EMC) The moisture content wood attains naturally at a given drybulb and wetbulb depression.

EVAPORATE Changing water from a liquid to a vapor form.

EXITING AIR CONTROL Sensing and controlling drybulb temperatures of heated kiln air as it leaves the lumber load.

FAN DECK The overhead area where the fans are mounted.

FAN PITCH The angle of the fan blades measured one-third of the fan radius in from the tip.

FAN SYSTEM The air moving equipment within a drykiln.

FANS, CROSS-SHAFT Fans with their shafts oriented perpendicular to the length of the kiln.

FANS, DIRECT-DRIVE Fans mounted directly on the fan motor shaft.

FANS, LINE-SHAFT Multiple fans mounted on a single shaft, which extends the length of the kiln. Used in conventional kilns.

FANS, SINGLE-SPEED Fans with motors that turn at one speed.

FANS, TWO-SPEED Fans with motors built with two sets of windings and capable of two rotation speeds.

FANS, VARIABLE-SPEED Fans whose motors turn at infinitely variable speeds by varying the line frequency of the power source.

FIBER SATURATION POINT (FSP) The moisture content at which moisture is saturated within the cell walls of wood and the cell cavities are free of water. This averages around 30 percent MC. Below FSP water is held in wood as bound water within the cell walls. Above FSP water is held as bound water and as free water within the cell cavities or lumen.

FINPIPE *See* COILS, STEAM.

FLITCHES A portion of a log sawed on two or more sides and intended for remanufacture.

FORKLIFT UNITS Lumber bundled or stacked into packages that can be lifted by a forklift.

FREE WATER Water held within the cell cavities of wood. Its movement is primarily by capillary action.

GRAIN The fibers in wood and their direction, size, arrangement, appearance, or quality.

GREEN LUMBER Lumber fresh from the saw.

HEARTWOOD Wood extending from the pith to the sapwood of a log or board. It is infiltrated with extractives, resins, and gums, which usually make it darker than the sapwood. Heartwood is very pronounced in southern pine.

HYDRAULIC To move or convey by fluid.

IMPLOSION The sudden collapse of kiln wall panels due to unequal air pressure between the inside and outside of the kiln. This usually occurs on startup as hot, expanded air suddenly cools and contracts as it passes through green lumber. The vents must open on startup to prevent this.

JUVENILE WOOD Wood deposited next to the pith and characterized by high longitudinal shrinkage.

KILN CARTS, TRUCKS A framework equipped with wheels that supports a load of lumber for kiln drying. Used in track kilns. Trucks also refer to the wheel assemblies alone.

KILN KINK The distortion in lumber that results from the uneven forces applied by misplaced truck supports and crossouts. Once dried into the lumber their effect is permanent.

KILNS, ACCELERATED Kilns that operate at temperatures above conventional temperatures (usually 200° F), but below high temperatures (above 212° F).

KILNS, CONVENTIONAL Kilns that operate at temperatures below 200° F.

KILNS, DIRECT-FIRED Kilns that are heated by blowing the products of combustion directly into the kiln.

KILNS, HIGH-TEMPERATURE Kilns that operate at temperatures above 212° F.

KILNS, PACKAGE A trackless compartment kiln for drying packages of stickered lumber. It has large doors through which a forklift loads lumber directly within the kiln.

KILNS, STEAM Kilns heated by steam.

KILNS, TRACK Kilns loaded by placing lumber on kiln trucks that run on tracks within the kiln. These can be single- or double-tracked kilns.

LATEWOOD Cellular growth that occurs after earlywood is deposited in the spring. Also called *summerwood*.

LIGNIN The thin, cementing layer between wood cells.

LONGITUDINAL DIRECTION Oriented in the direction of the length of the wood grain.

MANUFACTURE Any of the various processes that convert logs into lumber. *See also* EDGING, TRIMMING.

MISMANUFACTURE Wood imperfections or blemishes that result from manufacture. *See also* RABBETED EDGE, SNAKED SAWS, TAPER.

MODULATE To control within an infinite range between 0 percent and 100 percent as opposed to on/off control.

MOISTURE CONTENT Weight of the water contained in wood, expressed as a percentage of the weight of the ovendry wood.

$$\text{Moisture content} = \frac{(\text{Green weight} - \text{Ovendry weight})}{\text{Ovendry weight}} \times 100$$

MOISTURE DISTRIBUTION 1. Variation in moisture content within a board. 2. Variation in moisture content between boards.

MOISTURE GRADIENT The difference in moisture content between areas of a board. Usually refers to the moisture content difference between the surface and core of a board.

MULTIZONE Technique of dividing a kiln into more than two zones of heat control. Used in steam kilns.

OVERDRY Lumber dried to the point of having low moisture content.

PITCH 1. Amount of angle on a fan blade. 2. Accumulation of resin in the wood. This causes resistance moisture meters to read high. 3. To hurl horseshoes at a distant stake embedded in a sand pit.

PLC Programmable logic controllers/process logic controllers. Industrial computer with I/O capability, which programs in ladder-relay logic.

PLENUM BAFFLE A fixed or adjustable baffle suspended in the plenum chamber designed to split and direct airflow top to bottom through the lumber load.

PLENUM CHAMBER The space between the lumber and the outside walls of the kiln.

PNEUMATIC To move or convey an object by air pressure.

QUALITY CONTROL Any activity that helps to maximize the value of the raw material and lumber products through all phases of the manufacturing process. (T. D. Brown)

RABBETED EDGE Saw offset on the edge of lumber that is characteristic of Chip-n-Saws. Graded by minimum wane rules.

RADIAL A longitudinal surface or plane extending wholly or in part from the pith to the bark.

RAYS A ribbon-like aggregate of wood cells extending radially across the grain, radiating out from the pith to the bark.

REACTION WOOD Wood that forms in a tree in response to leaning. Called compression wood in softwoods, since it forms on the underside of the stem or limb. In hardwoods it forms on the upper side and is called tension wood.

RELATIVE HUMIDITY The ratio of the weight of water vapor in a given unit of air compared with the weight the same unit of air is capable of containing when fully saturated at the same temperature.

REVERSE CASEHARDENING A final stress-and-set condition in dry lumber in which the outer fibers are under tensile stress and the inner fibers are under compressive stress. This condition is not reversible. *See also* CASEHARDENING.

RIP MILL A manufacturing center that edges and trims lumber after drying to remove defects before grading and sorting for the planer mill.

RISERS Small distribution ducts within a direct-fired kiln that distribute heated air between the tracks from an underground main distribution duct.

RTD A temperature measuring device that measures the change in electrical resistance to determine temperature.

SAPWOOD The outer layers of a tree stem that contain living cells. The sapwood is distinctly lighter in southern pine than the heartwood it encloses.

SAW KERF The width of a saw cut as the saw passes through the log.

SCHEDULE The "recipe" for drying lumber, which dictates wetbulb and drybulb setpoints throughout the drying cycle.

SETPOINT The temperature to which you set your control instruments.

SHEAVES Pulleys mounted on the end of the drive and motor shaft over which belts pass to transfer rotational force from the motor to the shaft.

SHRINKAGE The contraction of wood to water loss below fiber saturation point. Expressed as a percentage of the green dimension.

SIGNAL A control impulse from a control device or sensor.

SILLS Wooden, concrete or steel supports placed on the ground or pavement onto which lumber is placed.

SNAKED SAWS Wavy saw line seen most often in band-sawed lumber. Graded as skipped dressing.

SOCK, WETBULB A fabric wick placed over a drybulb that cools the drybulb by evaporation to determine the relative humidity of the drying conditions.

SORTING Segregating rough-sawn lumber into lots for stacking, drying, and marketing purposes. Lumber is sorted for species, length, thickness, width, and occasionally moisture content.

SPECIES A group of plants or animals of a particular kind. For southern pine there are four major species (loblolly, longleaf, shortleaf, and slash) and three minor species (Virginia, pond, and spruce) of commercial interest.

SPIRAL GRAIN Characteristic of some trees where the wood fibers spiral around the tree rather than running straight. Causes slope of grain in lumber.

SPRINGWOOD *See* EARLYWOOD.

STACKER A mechanical device for making up unit packages of lumber for drying. As each layer of lumber is deposited on the unit, sticks are inserted manually or by machine between the layers.

STEAM Vapor produced by heating water.

STEAM SPRAYS Mechanical equipment for introducing water vapor or steam to the kiln environment.

STICKER A wooden, plastic, or metal strip laid between each layer of lumber as it is stacked for drying. Permits transfer of heat into and removal of water from the lumber stacks.

STICKLAYER The person who lays sticks.

SUMMERWOOD *See* LATEWOOD.

TANGENTIAL A surface plane that is perpendicular to the radial plane in wood.

TAPER Boards that are thicker or wider on one end than on the other. Graded as skip.

TARGET SIZES The sawn dimension of lumber that the mill endeavors to attain. Target sizes are composed of the finished lumber dimension plus allowances for planing, shrinkage, and variation in sawing accuracy to achieve skip-free lumber at the planing mill.

TEMPERATURE DROP ACROSS THE LOAD (TDAL) The difference in temperature between the air entering a package of lumber and its temperature as it leaves. This value is a measure of the heat demand of the lumber at that point and is an indirect measure of moisture content.

THERMAL OIL A heat transfer fluid used instead of steam. They are not interchangeable within the same kiln. Thermal oil is capable of higher temperatures at low operating pressure and is not susceptible to freezing.

THERMOCOUPLE A temperature sensing device composed of two dissimilar metals that produce an electromotive force (emf) that varies with temperature.

TRACHEIDS, LONGITUDINAL The elongated cells or fibers that comprise the majority of the anatomical structure of softwoods. Also present in some hardwoods.

TRACKS The steel rails on which kiln trucks rest and which serve to guide the lumber through the kiln for drying.

TRANSDUCER A device that converts power or a signal from one system to another. *See also* I/P TRANSDUCERS and P/I TRANSDUCERS.

TRANSDUCERS, I/P Convert electrical computer or control signals to air pressure signals to drive control valves and vents.

TRANSDUCERS, P/I Convert pressure signals to electrical signals for the computer or control device. These are used to measure steam pressure.

TRANSFER CARS A mechanical car for moving stacked units of lumber from one rail or track section to another.

TRAP KIT The internal parts of a trap that are replaced during servicing.

TRAP ORIFICE The internal opening within a trap through which steam is discharged.

TRAPS A device that separates condensate and air from the steam within the steam heating coils.

TRAPS, FLOAT AND THERMOSTATIC Traps that discharge condensate at a rate dependent on the position of an internal float. Air is removed through a thermostatic air vent.

TRAPS, INVERTED BUCKET Mechanical traps that operate on the difference in the density between steam and water.

TRAPS, ORIFICE Traps that discharge steam at a constant rate and have no internal moving parts to alter or stem its flow.

TRAPS, THERMOSTATIC Traps that use an internal bellows, which expands when steam flows to the trap, closing the orifice. As condensate accumulates the bellows contracts to release it.

TRIMMING The manufacturing step that squares the length of boards.

TRUCKS *See* KILN CARTS

TWIST A deviation flatwise, or a combination of flatwise and edgewise, in the form of a curl or spiral. *See* WARP.

VENTING The act of discharging hot, moisture-laden air from the kiln.

VENTING, CONVENTIONAL Venting by opening both rows of vents together.

VENTING, DEMAND Venting method by which the exhaust vents modulate open to maintain wetbulb setpoints and, failing this, the intake vents begin modulating open. Used in kilns with computerized controls.

VENTING, EXHAUST Venting by opening the exhaust vents only. Used in conventionally controlled high-temperature kilns.

VENTS The opening in the kiln roof or wall that can be opened and closed to control the wetbulb temperature in the kiln.

VENTS, EXHAUST Vents located on the pressure side of the fans.

VENTS, INTAKE Vents located on the low-pressure side of the fans.

WANE The common usage refers to bark edges on boards but also refers to the absence of wood on the edge of a board from any source.

WARP Any deviation from a true or plane surface in a board. *See also* BOW, CROOK, CUP, TWIST.

WATER BOX or TROUGH A water container mounted under the wetbulb that supplies water to the wick or sock.

WATER HAMMER Potentially damaging condition in steam pipes where waves develop on the condensate surface within a pipe as steam rushes over it. If the peak of the wave becomes high enough to reach the roof of the pipe, the water in the wave is pushed violently to the far end of the pipe.

WATERLOGGED Condition of steam coils where condensate is not discharged at a sufficient rate to keep it from accumulating in the coil. Filled with condensate, the coils do not transfer significant heat to the kiln.

WETBULB A drybulb with a wet sock pulled over it to measure the cooling effect of moisture evaporation from its surface. Used to monitor and maintain humidity conditions within the kiln.

WETBULB DEPRESSION The difference in wetbulb and drybulb temperatures.

ZONING The way a kiln is divided into temperature control areas.

REFERENCES

1. Brown, T. D. *Quality Control in Lumber Manufacture*. San Francisco: Miller Freeman Publications, 1982.
2. Forest Products Research Society, *Glossary of Terms Related to the Drying of Wood*. Madison, WI: FPRS, 1977.
3. Grant, E. L. *Statistical Quality Control*. New York: McGraw-Hill, 1964.
4. O'Dell, L. R. "Steam Traps—Basic Facts and Energy Saving Tips." *Heating/Piping/Air Conditioning 49*, No. 5 (May 1977).
5. Quasney, J. S. and J. Maniotes. *Standard BASIC Programming for Business and Management Application*. San Francisco: Boyd & Fraser, 1980.
6. Rasmussen, E. F. *Dry Kiln Operator's Manual*. Agricultural Handbook No. 188, Forest Service, GPO, 1961.
7. Southern Pine Inspection Bureau. *Grading Rules, 1977 Edition*. Pensacola, FL: SPIB, 1977.

Bibliography

Armstrong Machine Works. *Steam Conservation Guidelines for Condensate Drainage.* Three Rivers, MI: Armstrong Machine Works, 1976.

Bachrich, J. L. *Dry Kiln Handbook.* Vancouver, BC: H. A. Simons (International) Ltd., 1980.

Baker, H. D., *et al. Temperature Measurement in Engineering.* Vols. 1 and 2. Stamford, CT: Omega Press, 1975.

Brown, H. P., *et al. Textbook of Wood Technology.* Vols. 1 and 2. New York: McGraw-Hill, 1949.

Brown, T. D. *Quality Control in Lumber Manufacture.* San Francisco: Miller Freeman Publications, 1982.

Culpepper, L. G. "Application of Computerized Controls to Improve the Lumber Drying Operation." Seattle, WA: IUFRO Meeting, 1989.

———. "Coming Trends in Kiln Drying–What's New on the Horizon?" Reno, NV: West Coast Dry Kiln Association Meeting, May, 1988. Published in 1989 Proceedings.

Denig, J., *et al. Kiln Drying Southern Yellow Pine Lumber.* NCSU Agricultural Extension Service, Publication No. AG-402, 1988.

Forest Products Research Society, *Glossary of Terms Related to the Drying of Wood.* Madison, WI: FPRS, 1977.

Grant, E. L. *Statistical Quality Control.* New York: McGraw-Hill, 1964.

Hammond, H. "Principles of Operation to Be Considered in Selecting Steam Traps." *Plant Engineering.* Barrington, IL., July 1986, 70.

Hart, C. Arthur. "Kiln Overheat when Conditioning Lumber." *Forest Products Journal* 40, No. 4: 9–14.

Horton, Robert K. and Resch Helmuth. "Kiln Stick Thickness Effect on Drying 2-inch Western Hemlock Lumber." *Forest Products Journal* 26, No. 3: 35–40.

Irvington-Moore. *Instruction Manual for Round-Chart Recording Controllers, Models 4333-10W, 20W & 30W.* Jacksonville, FL., June 1980.

James, W. L., and R. B. Boone. "Capacitive In-Kiln Wood Moisture Content Monitors: Principles of Operation and Use." *Wood Science* (1982).

Koch, P. *Utilization of the Southern Pines.* Vols. 1 and 2. U.S. Department of Agriculture Forest Service, GPO Stock Number 0100-1483, 1972.

Makansi, J. *Managing Steam–An Engineering Guide to Commercial, Industrial and Utility Systems.* NJ: Leslie Company, 1985.

O'Dell, L. R. "Steam Traps–Basic Facts and Energy Saving Tips." *Heating/Piping/Air Conditioning* 49, No. 5 (May 1977).

Omega Engineering. *1987 Complete Temperature Measurement Handbook and Encyclopedia.* Stamford, CT: Omega Engineering, Inc., 1987.

Quasney, J. B. and J. Maniotes. *Standard BASIC Programming for Business and Management Applications*. San Francisco: Boyd & Fraser, 1980.

Rasmussen, E. F. *Dry Kiln Operator's Manual*. Agricultural Handbook No. 188, Forest Service, GPO, 1961.

Rosen, H. N. "Humidity Control Can Be Important When High Temperature Drying Hardwoods." *FPRB News-Digest*. Madison, WI: Drying and Storage Committee, 1977.

Sanders, C. M. *Inspection and Maintenance of High Temperature Dry Kilns and Equipment*. Jacksonville, FL: Irvington-Moore, 1986.

Southern Pine Inspection Bureau. *Grading Rules, 1977 Edition*. Pensacola, FL: SPIB, 1977.

Taylor, F. W. and P. H. Mitchell. "Can Kiln Drying Times Be Shortened Further?" *Forest Industries*. (November 1987).

Townsend, Kent. "Fire in the Dry Kiln." *FPRS Wood Drying News Digest*. Madison, WI: FPRS, 1981.

Wengert, E. M. "Trouble-shooting a Pneumatic Recorder/Controller." *FPRS News-Digest*. Madison, WI: Drying and Storage Committee, 1978.

Index

Boldface numbers refer to pages with *figures only* on the relevant subject.